THE

FARMER'S LAND-MEASURER,

OR

Pocket Companion;

SHOWING, AT ONE VIEW, THE

CONTENT OF ANY PIECE OF LAND,

FROM

DIMENSIONS TAKEN IN YARDS.

WITH A SET OF

USEFUL AGRICULTURAL TABLES

BY JAMES PEDDER,

EDITOR OF THE BOSTON CULTIVATOR.

NEW YORK:
C. M. SAXTON, AGRICULTURAL BOOK PUBLISHER.
1856.

ADVERTISEMENT.

The present work is offered to the Agricultural community, with perfect confidence. After the most rigid test of experience, it is found to be accurate in all its parts, and of the most simple application to all circumstances. It will prove itself an invaluable "Companion" to the real, practical man; giving him the information he seeks in comparatively an instant of time, without the labour of working over long sums; and its application can be made while engaged in the labours of the field, without previous preparation.

To the possessors of large tracts of land in the West, or open country, its use will soon create a *necessity* for its aid; for, to be enabled to set off any quantity of land in a field containing, perhaps, a hundred acres, in a few minutes of time, merely by stepping it and putting down four stakes; as also, to ascertain as quickly the quantity of land that has been ploughed, or planted, or cleared, out of the same tract; is an advantage, not to be appreciated until it shall have been enjoyed.

It is printed in the present form and size, that it might become "*The Farmer's Pocket Companion*" in reality.

C. M. SAXTON,

152 Fulton Street, New York.

PREFACE.

This work consists of Tables, so constructed as to give the content of any regular piece of land, measuring from one yard in length and breadth, to five hundred, by the addition of not more than three sums. And should the length or breadth exceed five hundred yards, its content may readily be found, by observing the rules which will be given in the course of the work. And to render the book as useful as possible, the preference is given to the use of *yards*, rather than to rods, or chains and links: as all farmers can tell the length and breadth of a piece of land by *pacing* or *stepping* it; therefore, in most cases, the content of a field might thus be known, without using rod or chain, sufficiently exact for paying labourers' wages; ascertaining the quantity of land ploughed in any given time; manuring, or apportioning seed at the time of sowing, as well as for harvesting the crops; without the necessity of employing a person to survey and measure — thus settling trifling disputes between masters and workmen, to the satisfaction of both, without calling in the aid of a third party.

There is added a table, which shows at one view what width is necessary to form a square of land of one acre, from one yard in length to five hundred. This is another mode of ascertaining the quantity of seed sown per acre; and the convenience of these tables will be very great, in showing the quantity of corn, grain, roots, &c., grown upon an acre; for it is only to step off, in any average part of the field, eleven yards square, and weigh or measure the produce, and forty times that quantity will be the exact product of an acre:—thus, if a bushel of barley, &c., be obtained from eleven yards square, the crop per acre will be ten bags of four bushels each, or forty bushels: then, by stepping the remainder of the field, and turning to the tables, the product of the whole field can be ascertained most easily. Again, if such a piece of land—namely, eleven yards square—produce three bags of potatoes, the produce per acre will be one hundred and twenty bags. Beets, turnips, carrots, parsneps, &c., may be weighed, and the quantity ascertained in a few minutes, by the same means. And when a farmer intends to plough an acre of land for his day's work, it is only to ascertain the *length* of the land by stepping, turn to the table to find the *width* necessary to form the acre, step that also, and place a rod at the spot; and he will derive much pleasure and interest from witnessing the progress which he is making towards the fulfilment of his task. This plan can be adopted at the time of harvest, creating great emulation amongst the persons employed, particularly if bands of workmen are engaged in different parts of the field; and in this way lightening materially their labour, by keeping alive a spirit of rivalry.

Added to these, are Tables for Manuring Land, showing how many loads will be required to cover an acre, the heaps being dropped at given distances, and the number of heaps in a load being first ascertained. A Table, showing the number of Plants required to plant an Acre. A Time Table, showing most accurately how to keep the time of any number of workmen, with the greatest ease and facility. Tables for Ploughing, &c., &c.; and a Table of the Measurement of the Bushel, showing when it is too large and when too small, according to the imperial standard of measurement; with many other things which will be found of great interest to the agriculturist.

JAMES PEDDER.
January, 1854.

CONTENTS.

THE

FARMER'S POCKET COMPANION.

METHOD OF MEASURING LAND.

THE content of land is estimated in acres, roods, and perches, forty of which perches make one rood, and four roods make one acre. In adding any different quantities of land into one sum, if the perches amount to more than *forty* and under *eighty,* set down the odd perches above forty, and carry *one* to the roods; if the perches exceed eighty and are under one hundred and twenty, set down the odd perches above eighty, and carry *two* to the roods; and so on, carrying *one* to the roods for every *forty perches,* and setting down the remainder under the perches. And if the roods, when added together, amount to more than four, carry *one* to the acre for every four roods, and set down the remainder under the roods.—The following examples will suffice for elucidation:—

First Example.			*Second Example.*			*Third Example.*		
A.	R.	P.	A.	R.	P.	A.	R.	P.
8	1	16	3	2	14	5	2	18
3	2	24	4	3	15	4	1	36
1	2	14	3	1	9	6	0	28
13	2	14	11	2	38	16	1	2

In the *first example,* the perches when added together amount to fifty-four; these being one rood and fourteen perches, I set down the fourteen under the perches, and carry *one* to the roods, which will make the roods *six,* or one acre two roods, which is the reason of the *two* being placed under the roods, the *one* being carried to the acres, which will then be found to be thirteen—say therefore, 13 acres 2 roods and 14 perches.

Second example—The perches amount to thirty-eight; these being less than forty, or one rood, I set down the thirty-eight under the perches, and carry *none* to the roods: the roods amounting to six, I set down *two* to the roods, and carry *one* to the acres, which will be found to be eleven—say therefore, 11 acres 2 roods and 38 perches.

Third example—The perches amount to eighty-two, I now set down two under the perches, and carry *two* to the roods, which will make the roods *five:* I therefore set down *one* under the roods, and carry *one* to the acres, which will then be sixteen—say therefore, 16 acres 1 rood and 2 perches.

It will now be shown how the content of various pieces of land and of different figures may be found by the following tables:—

When the piece to be measured is a square, like the annexed figure, measure the length and breadth in the middle, as there described.—Now, suppose the length from A to B to be 174 yards, breadth from C to D 87 yards, then turn to the column in the tables under 174 yards long, take out the sum which stands opposite 80 yards in width, and also out of the same column the sum which stands opposite 7 yards wide, and the two sums added together will give you the content of the field, namely, thus:—

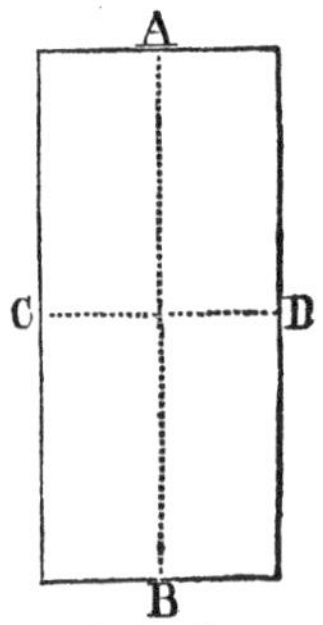

A
C D
B

	A.	R.	P.
174 yards long, 80 yards wide.........	2	3	20
174 yards long, 7 yards wide.........	0	1	0
Content,	3	0	20

Or, if a field of the *same figure* be 287 yards long and 154 yards wide, the content will be—

		A.	R.	P.
287 yards long,	100 yards wide.......	5	3	29
	50 do.	2	3	34
	4 do.	0	0	38
	Content,	9	0	21

Again—If the field be wider at one end than at the other, as in the figure here represented, measure the length in the middle, as in the first example, from A to B, and the breadth in the middle from C to D, and find the content as before. Now, suppose the length to be 303 yards, breadth 124 yards, the content will be—

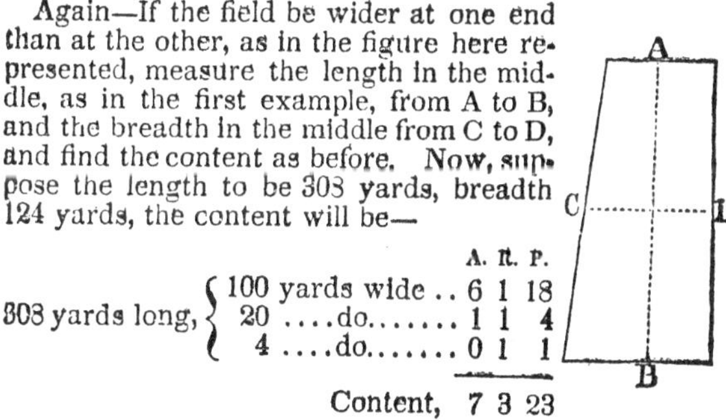

		A.	R.	P.
303 yards long,	100 yards wide ..	6	1	18
	20 ....do.......	1	1	4
	4 ....do.......	0	1	1
	Content,	7	3	23

But the breadth in the middle might have been found by measuring both ends and adding them together, and taking half their sum for a mean breadth—thus: If the larger end be 146 yards wide, and the lesser end be 102 yards wide, their sum would be 248 yards, the half of which is 124 yards for a mean breadth, as above; but the method above described is a shorter way to come at the result.

Once more — If one side of a piece of land be longer than the other, take the length and breadth exactly in the middle, and you will be able to find the content precisely as in the last example.— Now, suppose the length from A to B 318 yards, the breadth from C to D 129 yards—

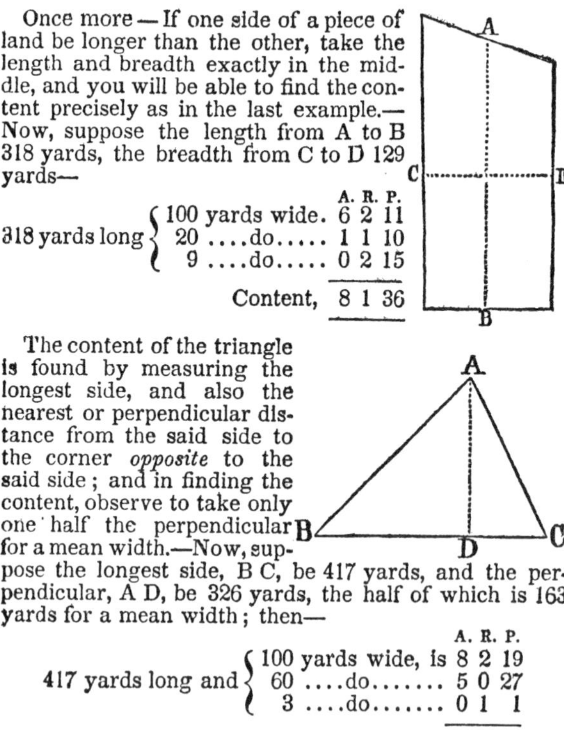

		A.	R.	P.
318 yards long	100 yards wide.	6	2	11
	20 ....do.....	1	1	10
	9 ....do.....	0	2	15
	Content,	8	1	36

The content of the triangle is found by measuring the longest side, and also the nearest or perpendicular distance from the said side to the corner *opposite* to the said side; and in finding the content, observe to take only one half the perpendicular for a mean width.—Now, suppose the longest side, B C, be 417 yards, and the perpendicular, A D, be 326 yards, the half of which is 163 yards for a mean width; then—

		A.	R.	P.
417 yards long and	100 yards wide, is	8	2	19
	60 ....do.......	5	0	27
	3 ....do.......	0	1	1
	Content of triangle,	14	0	7

And the content of any field, having four straight sides, may be found by the same rule:—measure the distance across the field from one longest corner to the other, which will throw it into two triangles; then measure the nearest or perpendicular distance from each of the other corners, to the line measured across the longest part of the field, and find the content of the two triangles as before: or, add the two perpendicular distances together, and take half the sum, which may be considered a mean width, and the line measured across the field being the length—thus:

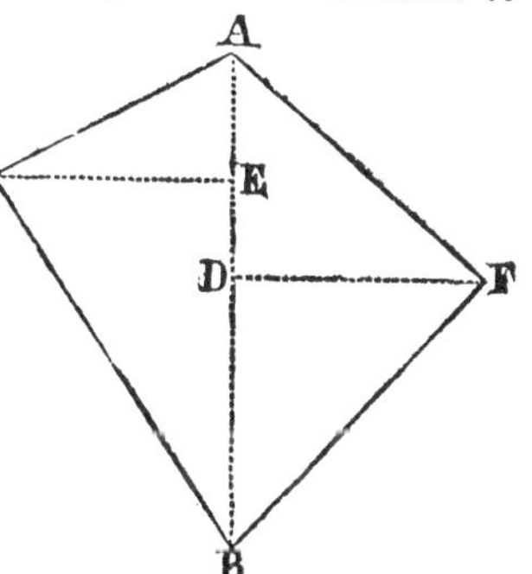

If the distance A B be 298 yards, the perpendicular distance C E 117 yards, and the perpendicular D F 113 yards, then 117 and 113 added together make 230, the half of which is 115 yards for a mean width; then—

		A.	R.	P.
298 yards long and	100 yards wide, is	6	0	25
	10 ....do.......	0	2	19
	5 ....do.......	0	1	9
	Content,	7	0	13

A field of any number of sides may be divided into triangles by measuring lines across it; and its content may then be found by the foregoing rules.

If the field have five sides, the lines A C and C E will divide it into three triangles, and its content is then found as follows:—Suppose A C 402 yards, B G 154, E F 162, E C 303, and D H 137—Now, to find the content of the parts A B, C E, we have B G 154, and E F 162, the sum of which is 306, and the half of this is 158 for a mean width; then—

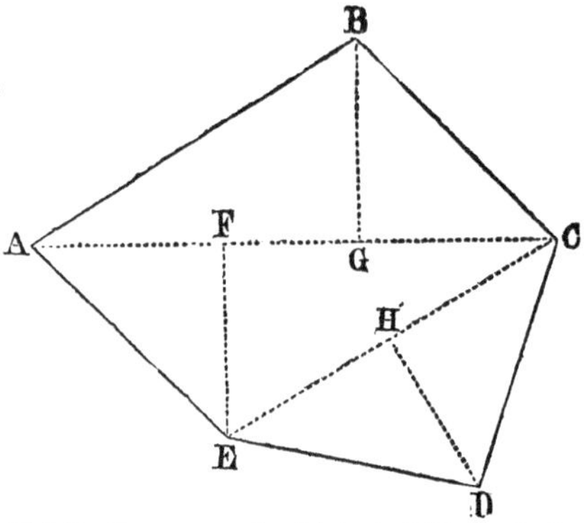

		A.	R.	P.
402 yards long and	100 yards wide, is	8	1	9
	50 ....do........	4	0	24
	8 ....do........	0	2	26
Content of the parts A B, C E,		13	0	9

And to find the content of the triangle E C D, we have D H 137 yards, the half of which is 68 yards for a mean width, and the length E C 303 yards—

	A.	R.	P.
303 yards long and 60 yards wide	3	3	1
...do........do... 8 ....do..........	0	2	0
Content of triangle E C D,	4	1	1
do. of the other part,	13	0	9
Whole content,	17	1	20

And in the same manner, a field of six, seven, or eight sides, may be divided into triangles by measuring lines across it; but to those unacquainted with land surveying, it would be best to take, if a ploughed field, a number of ridges at a time, and find their content by some of the first rules; then take more ridges and find their content also, until all the ridges are taken, and the summing up of the whole will be the content of the field.

Hitherto, the fences have always been considered as straight; but as it frequently happens that pieces of land are bounded by curved lines or fences, it is necessary to show how the content of such pieces can be found by these tables. And to do which, first set up a mark at each corner of the field; or, if the fence or line bend inward, then, as near the corner as possible, so as to be able to see the mark at the other end of the said fence or line, without being prevented by the curvature of the same: then find the content of the piece withinside of the said mark by some of the foregoing rules, and for the curved part measure the distance in a straight line from one mark to the other, against each of the curved fences—observing to set down the distance that your mark stands from the said fence where you first set off: and also note down the distance which the straight line you are measuring is from the fence or line, at every fifty yards distance in length; but if the fence be very irregular, take down the distance at every twenty or thirty yards in length, and also the distance at the end of the straight line; and in finding the content of the curve, set down every distance twice which was taken from the fence, except the distances taken at each end, which set down only *once;* then add them all together, and divide the sum by the number of lines added up, which gives the mean width of the curve, and the straight line measured against it is the length; thus—

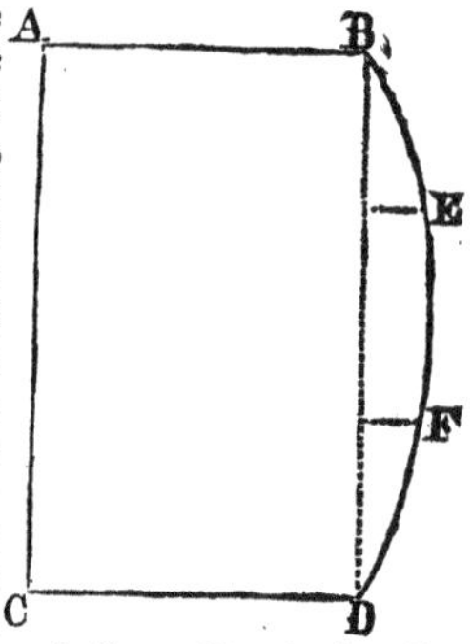

Suppose a field to be in the shape of the figure annexed, the length in the middle being 148 yards, the end A B 88 yards, the end C D 94 yards, and the straight line B D 154 yards; the distance from the straight line to the curve at B nothing, at E 18 yards, at F 13 yards, and at the end D, nothing: then, for the content of the square part the length is 148 yards, and the two ends when added together make 182 yards, the half of which is 91 for the mean width.—Again, for the curve, the length is 154; and the off-sets, by the aforesaid rule, will stand thus—

A B E F C D

```
    0
   18
   18
   13
   13
    0
 ----
 6)62
 ----
   10 the mean width of the curve.
```

		A.	R.	P.
148 yards long and	90 yards wide, is..	2	3	0
	1 ....do.........	0	0	5
154 yards long and	10 ....do.........	0	1	11
	Whole content,	3	0	16

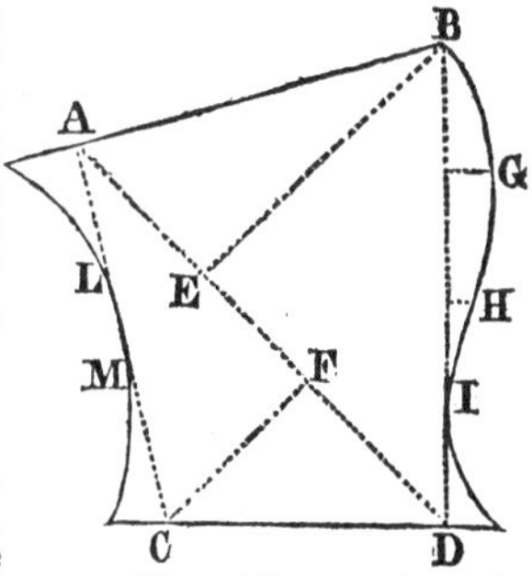

Again, suppose a piece of land of the following shape and dimensions:—A D 232 yards, B E 143 yards, C F 84 yards, and B D 200 yards; the distance from the straight line B D to the curve at B, nothing, at G 15 yards, at H 8 yards, at I nothing, and at the end D 14 yards; also, the line A C 160 yards long, and the distance from the end A of the straight line to the curve 23 yards, at L nothing, at M nothing, and at the end C 11 yards. From these dimensions, the content is found in the following manner:—The two perpendicular distances B E and C F, when added together, make 227, the half of which is 113 for a mean width of the square part: and to obtain a mean width of the curves, the offsets will stand in the following manner:—

0	
15	
15	23
8	0
8	0
0	0
0	0
14	11
8)60	6)34
Mean of curve B D, 7	Mean width of curve A C, 6

		A.	R.	P.
232 yards long and	100 yards wide, is	4	3	7
	10 ....do.......	0	1	37
	3 ....do.......	0	0	23
200 yards long and	7 ....do.......	0	1	6
160 ....do........	6 ....do.......	0	0	32
	Content of the whole field,	5	3	25

The content of various pieces of land of still more difficult measurement may be readily found by these tables; but as the present work is designed for the *practical Farmer*, and not for the *surveyor*, the subject need not be farther pursued.

If the length of a piece of land be more than 500 yards, the content may easily be found by the tables:—first marking off 500 yards in length, and adding the width, and then taking the remaining length above the 500 yards in length; and the sums added together will give the content of the whole field. Now, if for instance, the field be 738 yards long, and width 576, say—

		A.	R.	P.
500 yards long and	500 yards wide, is	51	2	24
	70 ....do......	7	0	37
	6 ....do......	0	2	19
238 yards long and	500 ....do......	24	2	14
	70 ....do......	3	1	31
	6 ....do......	0	1	7
	Content of the field,	87	3	12

To prove the truth of any of the following tables, multiply the length and breadth together, and divide the product by 4840—the number of square yards contained in an acre, and the quotient gives the number of acres; then multiply what remains by 4, and the product divided as before gives the roods; and the last remainder, if multiplied by 40 and divided in like manner, gives the perches.—Thus, by making use of the last example, where the length was 738 yards, and the width 576 yards—

```
           738
           576
         -----
          4428
         5166
        3690
        -------  A. R. P.
4840)425088(87  3  12
      38720
      -----
       37888
       33880
       -----
        4008
           4
       -----
       16032
       14520
       -----
        1512
          40
       -----
       60480
       4840
       -----
       12080
        9680
       -----
        2400
       -----
```

N. B. Where the parts of a perch do not amount to one-half, they are omitted in the following tables; but if they exceed one-half, then they are set down as a whole perch.

TABLES

FOR

FINDING THE CONTENT

OF ANY

PIECE OF LAND,

FROM

DIMENSIONS TAKEN IN YARDS.

From 1 to 5 Yards Long.

Width.	1 Yards Long.			2 Yards Long.			3 Yards Long.			4 Yards Long.			5 Yards Long.		
Yards.	*A.*	*R.*	*P.*	*A.*	*R.*	*P.*	*A.*	*R.*	*P.*	*A.*	*R.*	*P.*	*A.*	*R.*	*P.*
1	0	0	0	0	0	0	0	0	0	0	0	0	0	0	0
2	0	0	0	0	0	0	0	0	0	0	0	0	0	0	0
3	0	0	0	0	0	0	0	0	0	0	0	0	0	0	0
4	0	0	0	0	0	0	0	0	0	0	0	1	0	0	1
5	0	0	0	0	0	0	0	0	0	0	0	1	0	0	1
6	0	0	0	0	0	0	0	0	1	0	0	1	0	0	1
7	0	0	0	0	0	0	0	0	1	0	0	1	0	0	1
8	0	0	0	0	0	1	0	0	1	0	0	1	0	0	1
9	0	0	0	0	0	1	0	0	1	0	0	1	0	0	1
10	0	0	0	0	0	1	0	0	1	0	0	1	0	0	2
20	0	0	1	0	0	1	0	0	2	0	0	3	0	0	3
30	0	0	1	0	0	2	0	0	3	0	0	4	0	0	5
40	0	0	1	0	0	3	0	0	4	0	0	5	0	0	7
50	0	0	2	0	0	3	0	0	5	0	0	7	0	0	8
60	0	0	2	0	0	4	0	0	6	0	0	8	0	0	10
70	0	0	2	0	0	5	0	0	7	0	0	9	0	0	12
80	0	0	3	0	0	5	0	0	8	0	0	11	0	0	13
90	0	0	3	0	0	6	0	0	9	0	0	12	0	0	15
100	0	0	3	0	0	7	0	0	10	0	0	13	0	0	17
200	0	0	7	0	0	13	0	0	20	0	0	26	0	0	33
300	0	0	10	0	0	20	0	0	30	0	1	0	0	1	10
400	0	0	13	0	0	26	0	1	0	0	1	13	0	1	26
500	0	0	17	0	0	33	0	1	10	0	1	26	0	2	3

From 6 to 10 Yards Long.

Width.	6 Yards Long.			7 Yards Long.			8 Yards Long.			9 Yards Long.			10 Yards Long.		
Yards.	*A.*	*R.*	*P.*	*A.*	*R.*	*P.*	*A.*	*R.*	*P.*	*A.*	*R.*	*P.*	*A.*	*R.*	*P.*
1	0	0	0	0	0	0	0	0	0	0	0	0	0	0	0
2	0	0	0	0	0	0	0	0	1	0	0	1	0	0	1
3	0	0	1	0	0	1	0	0	1	0	0	1	0	0	1
4	0	0	1	0	0	1	0	0	1	0	0	1	0	0	1
5	0	0	1	0	0	1	0	0	1	0	0	1	0	0	2
6	0	0	1	0	0	1	0	0	2	0	0	2	0	0	2
7	0	0	1	0	0	2	0	0	2	0	0	2	0	0	2
8	0	0	2	0	0	2	0	0	2	0	0	2	0	0	3
9	0	0	2	0	0	2	0	0	2	0	0	3	0	0	3
10	0	0	2	0	0	2	0	0	3	0	0	3	0	0	3
20	0	0	4	0	0	5	0	0	5	0	0	6	0	0	7
30	0	0	6	0	0	7	0	0	8	0	0	9	0	0	10
40	0	0	8	0	0	9	0	0	11	0	0	12	0	0	13
50	0	0	10	0	0	12	0	0	13	0	0	15	0	0	17
60	0	0	12	0	0	14	0	0	16	0	0	18	0	0	20
70	0	0	14	0	0	16	0	0	19	0	0	21	0	0	23
80	0	0	16	0	0	19	0	0	21	0	0	24	0	0	26
90	0	0	18	0	0	21	0	0	24	0	0	27	0	0	30
100	0	0	20	0	0	23	0	0	26	0	0	30	0	0	33
200	0	1	0	0	1	6	0	1	13	0	1	20	0	1	26
300	0	1	20	0	1	29	0	1	39	0	2	0	0	2	19
400	0	1	39	0	2	13	0	2	26	0	2	39	0	3	12
500	0	2	19	0	2	36	0	3	12	0	3	29	1	0	5

From 11 to 15 Yards Long.

Width.	11 Yards Long.			12 Yards Long.			13 Yards Long.			14 Yards Long.			15 Yards Long.		
Yards.	*A.*	*R.*	*P.*	*A.*	*R.*	*P.*	*A.*	*R.*	*P.*	*A.*	*R.*	*P.*	*A.*	*R.*	*P.*
1	0	0	0	0	0	0	0	0	0	0	0	0	0	0	0
2	0	0	1	0	0	1	0	0	1	0	0	1	0	0	1
3	0	0	1	0	0	1	0	0	1	0	0	1	0	0	1
4	0	0	1	0	0	2	0	0	2	0	0	2	0	0	2
5	0	0	2	0	0	2	0	0	2	0	0	2	0	0	2
6	0	0	2	0	0	2	0	0	3	0	0	3	0	0	3
7	0	0	3	0	0	3	0	0	3	0	0	3	0	0	3
8	0	0	3	0	0	3	0	0	3	0	0	4	0	0	4
9	0	0	3	0	0	4	0	0	4	0	0	4	0	0	4
10	0	0	4	0	0	4	0	0	4	0	0	5	0	0	5
20	0	0	7	0	0	8	0	0	9	0	0	9	0	0	10
30	0	0	11	0	0	12	0	0	13	0	0	14	0	0	15
40	0	0	15	0	0	16	0	0	17	0	0	19	0	0	20
50	0	0	18	0	0	20	0	0	21	0	0	23	0	0	25
60	0	0	22	0	0	24	0	0	26	0	0	28	0	0	30
70	0	0	25	0	0	28	0	0	30	0	0	32	0	0	35
80	0	0	29	0	0	32	0	0	34	0	0	37	0	1	0
90	0	0	33	0	0	36	0	0	39	0	1	2	0	1	5
100	0	0	36	0	1	0	0	1	3	0	1	6	0	1	10
200	0	1	33	0	1	39	0	2	6	0	2	13	0	2	19
300	0	2	29	0	2	39	0	3	9	0	3	19	0	3	29
400	0	3	25	0	3	39	1	0	12	1	0	25	1	0	38
500	1	0	22	1	0	38	1	1	15	1	1	31	1	2	8

From 16 to 20 Yards Long.

Width.	16 Yards Long.			17 Yards Long.			18 Yards Long.			19 Yards Long.			20 Yards Long.		
Yards.	*A.*	*R.*	*P.*	*A.*	*R.*	*P.*	*A.*	*R.*	*P.*	*A.*	*R.*	*P.*	*A.*	*R.*	*P.*
1	0	0	1	0	0	1	0	0	1	0	0	1	0	0	1
2	0	0	1	0	0	1	0	0	1	0	0	1	0	0	1
3	0	0	2	0	0	2	0	0	2	0	0	2	0	0	2
4	0	0	2	0	0	2	0	0	2	0	0	3	0	0	3
5	0	0	3	0	0	3	0	0	3	0	0	3	0	0	3
6	0	0	3	0	0	3	0	0	4	0	0	4	0	0	4
7	0	0	4	0	0	4	0	0	4	0	0	4	0	0	5
8	0	0	4	0	0	4	0	0	5	0	0	5	0	0	5
9	0	0	5	0	0	5	0	0	5	0	0	6	0	0	6
10	0	0	5	0	0	6	0	0	6	0	0	6	0	0	7
20	0	0	11	0	0	11	0	0	12	0	0	13	0	0	13
30	0	0	16	0	0	17	0	0	18	0	0	19	0	0	20
40	0	0	21	0	0	22	0	0	24	0	0	25	0	0	26
50	0	0	26	0	0	28	0	0	30	0	0	31	0	0	33
60	0	0	32	0	0	34	0	0	36	0	0	38	0	1	0
70	0	0	37	0	0	39	0	1	2	0	1	4	0	1	6
80	0	1	2	0	1	5	0	1	8	0	1	10	0	1	13
90	0	1	8	0	1	11	0	1	14	0	1	17	0	1	20
100	0	1	13	0	1	16	0	1	20	0	1	23	0	1	26
200	0	2	26	0	2	32	0	2	39	0	3	6	0	3	12
300	0	3	39	1	0	9	1	0	19	1	0	28	1	0	38
400	1	1	12	1	1	25	1	1	38	1	2	11	1	2	24
500	1	2	24	1	3	1	1	3	18	1	3	34	2	0	11

From 21 to 25 Yards Long.

Width.	21 Yards Long.			22 Yards Long.			23 Yards Long.			24 Yards Long.			25 Yards Long.		
Yards.	*A.*	*R.*	*P.*	*A.*	*R.*	*P.*	*A.*	*R.*	*P.*	*A.*	*R.*	*P.*	*A.*	*R.*	*P.*
1	0	0	1	0	0	1	0	0	1	0	0	1	0	0	1
2	0	0	1	0	0	1	0	0	2	0	0	2	0	0	2
3	0	0	2	0	0	2	0	0	2	0	0	2	0	0	2
4	0	0	3	0	0	3	0	0	3	0	0	3	0	0	3
5	0	0	3	0	0	4	0	0	4	0	0	4	0	0	4
6	0	0	4	0	0	4	0	0	5	0	0	5	0	0	5
7	0	0	5	0	0	5	0	0	5	0	0	6	0	0	6
8	0	0	6	0	0	6	0	0	6	0	0	6	0	0	7
9	0	0	6	0	0	7	0	0	7	0	0	7	0	0	7
10	0	0	7	0	0	7	0	0	8	0	0	8	0	0	8
20	0	0	14	0	0	15	0	0	15	0	0	16	0	0	17
30	0	0	21	0	0	22	0	0	23	0	0	24	0	0	25
40	0	0	28	0	0	29	0	0	30	0	0	32	0	0	33
50	0	0	35	0	0	36	0	0	38	0	1	0	0	1	1
60	0	1	2	0	1	4	0	1	6	0	1	8	0	1	10
70	0	1	9	0	1	11	0	1	13	0	1	16	0	1	18
80	0	1	16	0	1	18	0	1	21	0	1	23	0	1	26
90	0	1	22	0	1	25	0	1	28	0	1	31	0	1	34
100	0	1	29	0	1	33	0	1	36	0	1	39	0	2	3
200	0	3	19	0	3	25	0	3	32	0	3	39	1	0	5
300	1	1	8	1	1	18	1	1	28	1	1	38	1	2	8
400	1	2	38	1	3	11	1	3	24	1	3	37	2	0	11
500	2	0	27	2	1	4	2	1	20	2	1	37	2	2	13

From 26 to 30 Yards Long.

Width.	26 Yards Long.			27 Yards Long.			28 Yards Long.			29 Yards Long.			30 Yards Long.		
Yards.	*A.*	*R.*	*P.*	*A.*	*R.*	*P.*	*A.*	*R.*	*P.*	*A.*	*R.*	*P.*	*A.*	*R.*	*P.*
1	0	0	1	0	0	1	0	0	1	0	0	1	0	0	1
2	0	0	2	0	0	2	0	0	2	0	0	2	0	0	2
3	0	0	3	0	0	3	0	0	3	0	0	3	0	0	3
4	0	0	3	0	0	4	0	0	4	0	0	4	0	0	4
5	0	0	4	0	0	4	0	0	5	0	0	5	0	0	5
6	0	0	5	0	0	5	0	0	6	0	0	6	0	0	6
7	0	0	6	0	0	6	0	0	6	0	0	7	0	0	7
8	0	0	7	0	0	7	0	0	7	0	0	8	0	0	8
9	0	0	8	0	0	8	0	0	8	0	0	9	0	0	9
10	0	0	9	0	0	9	0	0	9	0	0	10	0	0	10
20	0	0	17	0	0	18	0	0	19	0	0	19	0	0	20
30	0	0	26	0	0	27	0	0	28	0	0	29	0	0	30
40	0	0	34	0	0	36	0	0	37	0	0	38	0	1	0
50	0	1	3	0	1	5	0	1	6	0	1	8	0	1	10
60	0	1	12	0	1	14	0	1	16	0	1	18	0	1	20
70	0	1	20	0	1	22	0	1	25	0	1	27	0	1	29
80	0	1	29	0	1	31	0	1	34	0	1	37	0	1	39
90	0	1	37	0	2	0	0	2	3	0	2	6	0	2	9
100	0	2	6	0	2	9	0	2	13	0	2	16	0	2	19
200	1	0	12	1	0	19	1	0	25	1	0	32	1	0	38
300	1	2	18	1	2	28	1	2	38	1	3	8	1	3	18
400	2	0	24	2	0	37	2	1	10	2	1	23	2	1	37
500	2	2	30	2	3	6	2	3	23	2	3	39	3	0	16

From 31 to 35 Yards Long.

Width.	31 Yards Long.			32 Yards Long.			33 Yards Long.			34 Yards Long.			35 Yards Long.		
Yards.	*A.*	*R.*	*P.*	*A.*	*R.*	*P.*	*A.*	*R.*	*P.*	*A.*	*R.*	*P.*	*A.*	*R.*	*P.*
1	0	0	1	0	0	1	0	0	1	0	0	1	0	0	1
2	0	0	2	0	0	2	0	0	2	0	0	2	0	0	2
3	0	0	3	0	0	3	0	0	3	0	0	3	0	0	3
4	0	0	4	0	0	4	0	0	4	0	0	4	0	0	5
5	0	0	5	0	0	5	0	0	5	0	0	6	0	0	6
6	0	0	6	0	0	6	0	0	7	0	0	7	0	0	7
7	0	0	7	0	0	7	0	0	8	0	0	8	0	0	8
8	0	0	8	0	0	8	0	0	9	0	0	9	0	0	9
9	0	0	9	0	0	10	0	0	10	0	0	10	0	0	10
10	0	0	10	0	0	11	0	0	11	0	0	11	0	0	12
20	0	0	20	0	0	21	0	0	22	0	0	22	0	0	23
30	0	0	31	0	0	32	0	0	33	0	0	34	0	0	35
40	0	1	1	0	1	2	0	1	4	0	1	5	0	1	6
50	0	1	11	0	1	13	0	1	15	0	1	16	0	1	18
60	0	1	21	0	1	23	0	1	25	0	1	27	0	1	29
70	0	1	32	0	1	34	0	1	36	0	1	39	0	2	1
80	0	2	2	0	2	5	0	2	7	0	2	10	0	2	13
90	0	2	12	0	2	15	0	2	18	0	2	21	0	2	24
100	0	2	22	0	2	26	0	2	29	0	2	32	0	2	36
200	1	1	5	1	1	12	1	1	18	1	1	25	1	1	31
300	1	3	27	1	3	37	2	0	7	2	0	17	2	0	27
400	2	2	10	2	2	23	2	2	36	2	3	10	2	3	33
500	3	0	32	3	1	9	3	1	25	3	2	2	3	2	19

From 36 to 40 Yards Long.

Width.	36 Yards Long.			37 Yards Long.			38 Yards Long.			39 Yards Long.			40 Yards Long.		
Yards.	*A.*	*R.*	*P.*	*A.*	*R.*	*P.*	*A.*	*R.*	*P.*	*A.*	*R.*	*P.*	*A.*	*R.*	*P.*
1	0	0	1	0	0	1	0	0	1	0	0	1	0	0	1
2	0	0	2	0	0	2	0	0	3	0	0	3	0	0	3
3	0	0	4	0	0	4	0	0	4	0	0	4	0	0	4
4	0	0	5	0	0	5	0	0	5	0	0	5	0	0	5
5	0	0	6	0	0	6	0	0	6	0	0	6	0	0	7
6	0	0	7	0	0	7	0	0	8	0	0	8	0	0	8
7	0	0	8	0	0	9	0	0	9	0	0	9	0	0	9
8	0	0	10	0	0	10	0	0	10	0	0	10	0	0	11
9	0	0	11	0	0	11	0	0	11	0	0	12	0	0	12
10	0	0	12	0	0	12	0	0	13	0	0	13	0	0	13
20	0	0	24	0	0	24	0	0	25	0	0	26	0	0	26
30	0	0	36	0	0	37	0	0	38	0	0	39	0	1	0
40	0	1	8	0	1	9	0	1	10	0	1	12	0	1	13
50	0	1	20	0	1	21	0	1	23	0	1	24	0	1	26
60	0	1	31	0	1	33	0	1	35	0	1	37	0	1	39
70	0	2	3	0	2	6	0	2	8	0	2	10	0	2	13
80	0	2	15	0	2	18	0	2	20	0	2	23	0	2	26
90	0	2	27	0	2	30	0	2	33	0	2	36	0	2	39
100	0	2	39	0	3	2	0	3	6	0	3	9	0	3	12
200	1	1	38	1	2	5	1	2	11	1	2	18	1	2	24
300	2	0	37	2	1	7	2	1	17	2	1	27	2	1	37
400	2	3	36	3	0	9	3	0	22	3	0	36	3	0	9
500	3	2	35	3	3	12	3	3	28	4	0	5	4	1	21

From 41 to 45 Yards Long.

Width.	41 Yards Long.			42 Yards Long.			43 Yards Long.			44 Yards Long.			45 Yards Long.		
Yards.	*A.*	*R.*	*P.*	*A.*	*R.*	*P.*	*A.*	*R.*	*P.*	*A.*	*R.*	*P.*	*A.*	*R.*	*P.*
1	0	0	1	0	0	1	0	0	1	0	0	1	0	0	1
2	0	0	3	0	0	3	0	0	3	0	0	3	0	0	3
3	0	0	4	0	0	4	0	0	4	0	0	4	0	0	4
4	0	0	6	0	0	6	0	0	6	0	0	6	0	0	6
5	0	0	7	0	0	7	0	0	7	0	0	7	0	0	7
6	0	0	8	0	0	8	0	0	9	0	0	9	0	0	9
7	0	0	9	0	0	10	0	0	10	0	0	10	0	0	10
8	0	0	11	0	0	11	0	0	11	0	0	12	0	0	12
9	0	0	12	0	0	12	0	0	13	0	0	13	0	0	13
10	0	0	14	0	0	14	0	0	14	0	0	15	0	0	15
20	0	0	27	0	0	28	0	0	28	0	0	29	0	0	30
30	0	1	1	0	1	2	0	1	3	0	1	4	0	1	5
40	0	1	14	0	1	16	0	1	17	0	1	18	0	1	20
50	0	1	28	0	1	29	0	1	31	0	1	33	0	1	34
60	0	2	1	0	2	3	0	2	5	0	2	7	0	2	9
70	0	2	15	0	2	17	0	2	20	0	2	22	0	2	24
80	0	2	28	0	2	31	0	2	34	0	2	36	0	2	39
90	0	3	2	0	3	5	0	3	8	0	3	11	0	3	14
100	0	3	16	0	3	19	0	3	22	0	3	25	0	3	29
200	1	2	31	1	2	38	1	3	4	1	3	11	1	3	18
300	2	2	7	2	2	17	2	2	26	2	2	36	2	3	6
400	3	1	22	3	1	35	3	2	9	3	2	22	3	2	35
500	4	0	38	4	1	14	4	1	31	4	2	7	4	2	24

From 46 to 50 Yards Long.

Width.	46 Yards Long.			47 Yards Long.			48 Yards Long.			49 Yards Long.			50 Yards Long.		
Yards.	*A.*	*R.*	*P.*	*A.*	*R.*	*P.*	*A.*	*R.*	*P.*	*A.*	*R.*	*P.*	*A.*	*R.*	*P.*
1	0	0	2	0	0	2	0	0	2	0	0	2	0	0	2
2	0	0	3	0	0	3	0	0	3	0	0	3	0	0	3
3	0	0	5	0	0	5	0	0	5	0	0	5	0	0	5
4	0	0	6	0	0	6	0	0	6	0	0	6	0	0	7
5	0	0	8	0	0	8	0	0	8	0	0	8	0	0	8
6	0	0	9	0	0	9	0	0	10	0	0	10	0	0	10
7	0	0	11	0	0	11	0	0	11	0	0	11	0	0	12
8	0	0	12	0	0	12	0	0	13	0	0	13	0	0	13
9	0	0	14	0	0	14	0	0	14	0	0	15	0	0	15
10	0	0	15	0	0	16	0	0	16	0	0	16	0	0	17
20	0	0	30	0	0	31	0	0	32	0	0	32	0	0	33
30	0	1	6	0	1	7	0	1	8	0	1	9	0	1	10
40	0	1	21	0	1	22	0	1	23	0	1	25	0	1	26
50	0	1	36	0	1	38	0	1	39	0	2	1	0	2	3
60	0	2	11	0	2	13	0	2	15	0	2	17	0	2	19
70	0	2	26	0	2	29	0	2	31	0	2	33	0	2	36
80	0	3	2	0	3	4	0	3	7	0	3	10	0	3	12
90	0	3	17	0	3	20	0	3	23	0	3	26	0	3	29
100	0	3	32	0	3	35	0	3	39	1	0	2	1	0	5
200	1	3	24	1	3	31	1	3	37	2	0	4	2	0	11
300	2	3	16	2	3	26	2	3	36	3	0	6	3	0	16
400	3	3	8	3	3	21	3	3	35	4	0	8	4	0	21
500	4	3	0	4	3	17	4	3	33	5	0	10	5	0	26

From 51 to 55 Yards Long.

Width.	51 Yards Long.			52 Yards Long.			53 Yards Long.			54 Yards Long.			55 Yards Long.		
Yards.	*A.*	*R.*	*P.*	*A.*	*R.*	*P.*	*A.*	*R.*	*P.*	*A.*	*R.*	*P.*	*A.*	*R.*	*P.*
1	0	0	2	0	0	2	0	0	2	0	0	2	0	0	2
2	0	0	3	0	0	3	0	0	4	0	0	4	0	0	4
3	0	0	5	0	0	5	0	0	5	0	0	5	0	0	5
4	0	0	7	0	0	7	0	0	7	0	0	7	0	0	7
5	0	0	8	0	0	9	0	0	9	0	0	9	0	0	9
6	0	0	10	0	0	10	0	0	11	0	0	11	0	0	11
7	0	0	12	0	0	12	0	0	12	0	0	12	0	0	13
8	0	0	13	0	0	14	0	0	14	0	0	14	0	0	15
9	0	0	15	0	0	15	0	0	16	0	0	16	0	0	16
10	0	0	17	0	0	17	0	0	18	0	0	18	0	0	18
20	0	0	34	0	0	34	0	0	35	0	0	36	0	0	36
30	0	1	11	0	1	12	0	1	13	0	1	14	0	1	15
40	0	1	27	0	1	29	0	1	30	0	1	31	0	1	33
50	0	2	4	0	2	6	0	2	8	0	2	9	0	2	11
60	0	2	21	0	2	23	0	2	25	0	2	27	0	2	29
70	0	2	38	0	3	0	0	3	3	0	3	5	0	3	7
80	0	3	15	0	3	18	0	3	20	0	3	23	0	3	25
90	0	3	32	0	3	35	0	3	38	1	0	1	1	0	4
100	1	0	9	1	0	12	1	0	15	1	0	19	1	0	22
200	2	0	17	2	0	24	2	0	30	2	0	37	2	1	5
300	3	0	26	3	0	36	3	1	6	3	1	16	3	1	25
400	4	0	34	4	1	8	4	1	21	4	1	34	4	2	7
500	5	1	3	5	1	20	5	1	36	5	2	13	5	2	29

From 56 to 60 Yards Long.

Width.	56 Yards Long.			57 Yards Long.			58 Yards Long.			59 Yards Long.			60 Yards Long.		
Yards.	*A.*	*R.*	*P.*	*A.*	*R.*	*P.*	*A.*	*R.*	*P.*	*A.*	*R.*	*P.*	*A.*	*R.*	*P.*
1	0	0	2	0	0	2	0	0	2	0	0	2	0	0	2
2	0	0	4	0	0	4	0	0	4	0	0	4	0	0	4
3	0	0	6	0	0	6	0	0	6	0	0	6	0	0	6
4	0	0	7	0	0	8	0	0	8	0	0	8	0	0	8
5	0	0	9	0	0	9	0	0	10	0	0	10	0	0	10
6	0	0	11	0	0	11	0	0	12	0	0	12	0	0	12
7	0	0	13	0	0	13	0	0	13	0	0	14	0	0	14
8	0	0	15	0	0	15	0	0	15	0	0	16	0	0	16
9	0	0	17	0	0	17	0	0	17	0	0	18	0	0	18
10	0	0	19	0	0	19	0	0	19	0	0	20	0	0	20
20	0	0	37	0	0	38	0	0	38	0	0	39	0	1	0
30	0	1	16	0	1	17	0	1	18	0	1	19	0	1	20
40	0	1	34	0	1	35	0	1	37	0	1	38	0	1	39
50	0	2	13	0	2	14	0	2	16	0	2	18	0	2	19
60	0	2	31	0	2	33	0	2	35	0	2	37	0	2	39
70	0	3	10	0	3	12	0	3	14	0	3	17	0	3	19
80	0	3	28	0	3	31	0	3	33	0	3	36	0	3	39
90	1	0	7	1	0	10	1	0	13	1	0	16	1	0	19
100	1	0	25	1	0	28	1	0	32	1	0	35	1	0	38
200	2	1	10	2	1	17	2	1	23	2	1	30	2	1	37
300	3	1	35	3	2	5	3	2	15	3	2	25	3	2	35
400	4	2	20	4	2	34	4	3	7	4	3	20	4	3	33
500	5	3	6	5	3	22	5	3	39	6	0	15	6	0	32

From 61 to 65 Yards Long.

Width.	61 Yards Long.			62 Yards Long.			63 Yards Long.			64 Yards Long.			65 Yards Long.		
Yards.	*A.*	*R.*	*P.*	*A.*	*R.*	*P.*	*A.*	*R.*	*P.*	*A.*	*R.*	*P.*	*A.*	*R.*	*P.*
1	0	0	2	0	0	2	0	0	2	0	0	2	0	0	2
2	0	0	4	0	0	4	0	0	4	0	0	4	0	0	4
3	0	0	6	0	0	6	0	0	6	0	0	6	0	0	6
4	0	0	8	0	0	8	0	0	8	0	0	8	0	0	9
5	0	0	10	0	0	10	0	0	10	0	0	11	0	0	11
6	0	0	12	0	0	12	0	0	12	0	0	13	0	0	13
7	0	0	14	0	0	14	0	0	15	0	0	15	0	0	15
8	0	0	16	0	0	16	0	0	17	0	0	17	0	0	17
9	0	0	18	0	0	18	0	0	19	0	0	19	0	0	19
10	0	0	20	0	0	20	0	0	21	0	0	21	0	0	21
20	0	1	0	0	1	1	0	1	2	0	1	2	0	1	3
30	0	1	20	0	1	21	0	1	22	0	1	23	0	1	24
40	0	2	1	0	2	2	0	2	3	0	2	5	0	2	6
50	0	2	21	0	2	22	0	2	24	0	2	26	0	2	27
60	0	3	1	0	3	3	0	3	5	0	3	7	0	3	9
70	0	3	21	0	3	23	0	3	26	0	3	28	0	3	30
80	1	0	1	1	0	4	1	0	7	1	0	9	1	0	12
90	1	0	21	1	0	24	1	0	27	1	0	30	1	0	33
100	1	1	2	1	1	5	1	1	8	1	1	12	1	1	15
200	2	2	3	2	2	10	2	2	17	2	2	23	2	2	30
300	3	3	5	3	3	15	3	3	25	3	3	35	4	0	5
400	5	0	7	5	0	20	5	0	33	5	1	6	5	1	20
500	6	1	8	6	1	25	6	2	1	6	2	18	6	2	34

From 66 to 70 Yards Long.

Width.	66 Yards Long.			67 Yards Long.			68 Yards Long.			69 Yards Long.			70 Yards Long.		
Yards.	*A.*	*R.*	*P.*	*A.*	*R.*	*P.*	*A.*	*R.*	*P.*	*A.*	*R.*	*P.*	*A.*	*R.*	*P.*
1	0	0	2	0	0	2	0	0	2	0	0	2	0	0	2
2	0	0	4	0	0	4	0	0	4	0	0	5	0	0	5
3	0	0	7	0	0	7	0	0	7	0	0	7	0	0	7
4	0	0	9	0	0	9	0	0	9	0	0	9	0	0	9
5	0	0	11	0	0	11	0	0	11	0	0	11	0	0	12
6	0	0	13	0	0	13	0	0	13	0	0	14	0	0	14
7	0	0	15	0	0	16	0	0	16	0	0	16	0	0	16
8	0	0	17	0	0	18	0	0	18	0	0	18	0	0	19
9	0	0	20	0	0	20	0	0	20	0	0	21	0	0	21
10	0	0	22	0	0	22	0	0	22	0	0	23	0	0	23
20	0	1	4	0	1	4	0	1	5	0	1	6	0	1	6
30	0	1	25	0	1	26	0	1	27	0	1	28	0	1	29
40	0	2	7	0	2	9	0	2	10	0	2	11	0	2	13
50	0	2	29	0	2	31	0	2	32	0	2	34	0	2	36
60	0	3	11	0	3	13	0	3	15	0	3	17	0	3	19
70	0	3	33	0	3	35	0	3	37	1	0	0	1	0	2
80	1	0	15	1	0	17	1	0	20	1	0	22	1	0	25
90	1	0	36	1	0	39	1	1	2	1	1	5	1	1	8
100	1	1	18	1	1	21	1	1	25	1	1	28	1	1	31
200	2	2	36	2	3	3	2	3	10	2	3	16	2	3	23
300	4	0	15	4	0	24	4	0	34	4	1	4	4	1	14
400	5	1	33	5	2	6	5	2	19	5	2	32	5	3	6
500	6	3	11	6	3	27	7	0	4	7	0	20	7	0	37

From 71 to 75 Yards Long.

Width.	71 Yards Long.			72 Yards Long.			73 Yards Long.			74 Yards Long.			75 Yards Long.		
Yards.	*A.*	*R.*	*P.*	*A.*	*R.*	*P.*	*A.*	*R.*	*P.*	*A.*	*R.*	*P.*	*A.*	*R.*	*P.*
1	0	0	2	0	0	2	0	0	2	0	0	2	0	0	2
2	0	0	5	0	0	5	0	0	5	0	0	5	0	0	5
3	0	0	7	0	0	7	0	0	7	0	0	7	0	0	7
4	0	0	9	0	0	10	0	0	10	0	0	10	0	0	10
5	0	0	12	0	0	12	0	0	12	0	0	12	0	0	12
6	0	0	14	0	0	14	0	0	14	0	0	15	0	0	15
7	0	0	16	0	0	17	0	0	17	0	0	17	0	0	17
8	0	0	19	0	0	19	0	0	19	0	0	20	0	0	20
9	0	0	21	0	0	21	0	0	22	0	0	22	0	0	22
10	0	0	23	0	0	24	0	0	24	0	0	24	0	0	25
20	0	1	7	0	1	8	0	1	8	0	1	9	0	1	10
30	0	1	30	0	1	31	0	1	32	0	1	33	0	1	34
40	0	2	14	0	2	15	0	2	17	0	2	18	0	2	19
50	0	2	37	0	2	39	0	3	1	0	3	2	0	3	4
60	0	3	21	0	3	23	0	3	25	0	3	27	0	3	29
70	1	0	4	1	0	7	1	0	9	1	0	11	1	0	14
80	1	0	28	1	0	30	1	0	33	1	0	36	1	0	38
90	1	1	11	1	1	14	1	1	17	1	1	20	1	1	23
100	1	1	35	1	1	38	1	2	1	1	2	5	1	2	8
200	2	3	29	2	3	36	3	0	3	3	0	9	3	0	16
300	4	1	24	4	1	34	4	2	4	4	2	14	4	2	24
400	5	3	19	5	3	32	6	0	5	6	0	19	6	0	32
500	7	1	14	7	1	30	7	2	7	7	2	23	7	3	0

From 76 to 80 Yards Long.

Width.	76 Yards Long.			77 Yards Long.			78 Yards Long.			79 Yards Long.			80 Yards Long.		
Yards.	*A.*	*R.*	*P.*	*A.*	*R.*	*P.*	*A.*	*R.*	*P.*	*A.*	*R.*	*P.*	*A.*	*R.*	*P.*
1	0	0	3	0	0	3	0	0	3	0	0	3	0	0	3
2	0	0	5	0	0	5	0	0	5	0	0	5	0	0	5
3	0	0	8	0	0	8	0	0	8	0	0	8	0	0	8
4	0	0	10	0	0	10	0	0	10	0	0	10	0	0	11
5	0	0	13	0	0	13	0	0	13	0	0	13	0	0	13
6	0	0	15	0	0	15	0	0	15	0	0	16	0	0	16
7	0	0	18	0	0	18	0	0	18	0	0	18	0	0	19
8	0	0	20	0	0	20	0	0	21	0	0	21	0	0	21
9	0	0	23	0	0	23	0	0	23	0	0	24	0	0	24
10	0	0	25	0	0	25	0	0	26	0	0	26	0	0	26
20	0	1	10	0	1	11	0	1	12	0	1	12	0	1	13
30	0	1	35	0	1	36	0	1	37	0	1	38	0	1	39
40	0	2	20	0	2	22	0	2	23	0	2	24	0	2	26
50	0	3	6	0	3	7	0	3	9	0	3	11	0	3	12
60	0	3	31	0	3	33	0	3	35	0	3	37	0	3	39
70	1	0	16	1	0	18	1	0	20	1	0	23	1	0	25
80	1	1	1	1	1	4	1	1	6	1	1	9	1	1	12
90	1	1	26	1	1	29	1	1	32	1	1	35	1	1	38
100	1	2	11	1	2	15	1	2	18	1	2	21	1	2	24
200	3	0	22	3	0	29	3	0	36	3	1	2	3	1	9
300	4	2	34	4	3	4	4	3	14	4	3	23	4	3	33
400	6	1	5	6	1	18	6	1	31	6	2	5	6	2	18
500	7	3	16	7	3	33	8	0	9	8	0	26	8	1	2

From 81 to 85 Yards Long.

Width.	81 Yards Long.			82 Yards Long			83 Yards Long.			84 Yards Long.			85 Yards Long.		
Yards.	*A.*	*R.*	*P.*	*A.*	*R.*	*P.*	*A.*	*R.*	*P.*	*A.*	*R.*	*P.*	*A.*	*R.*	*P.*
1	0	0	3	0	0	3	0	0	3	0	0	3	0	0	3
2	0	0	5	0	0	5	0	0	5	0	0	6	0	0	6
3	0	0	8	0	0	8	0	0	8	0	0	8	0	0	8
4	0	0	11	0	0	11	0	0	11	0	0	11	0	0	11
5	0	0	13	0	0	14	0	0	14	0	0	14	0	0	14
6	0	0	16	0	0	16	0	0	16	0	0	17	0	0	17
7	0	0	19	0	0	19	0	0	19	0	0	19	0	0	20
8	0	0	21	0	0	22	0	0	22	0	0	22	0	0	22
9	0	0	24	0	0	24	0	0	25	0	0	25	0	0	25
10	0	0	27	0	0	27	0	0	27	0	0	28	0	0	28
20	0	1	14	0	1	14	0	1	15	0	1	16	0	1	16
30	0	2	0	0	2	1	0	2	2	0	2	3	0	2	4
40	0	2	27	0	2	28	0	2	30	0	2	31	0	2	32
50	0	3	14	0	3	16	0	3	17	0	3	19	0	3	20
60	1	0	1	1	0	3	1	0	5	1	0	7	1	0	9
70	1	0	27	1	0	30	1	0	32	1	0	34	1	0	37
80	1	1	14	1	1	17	1	1	20	1	1	22	1	1	25
90	1	2	1	1	2	4	1	2	7	1	2	10	1	2	13
100	1	2	28	1	2	31	1	2	34	1	2	38	1	3	1
200	3	1	16	3	1	22	3	1	29	3	1	35	3	2	2
300	5	0	3	5	0	13	5	0	23	5	0	33	5	1	3
400	6	2	31	6	3	4	6	3	18	6	3	31	7	0	4
500	8	1	19	8	1	35	8	2	12	8	2	28	8	3	5

From 86 to 90 Yards Long.

Width.	86 Yards Long.			87 Yards Long.			88 Yards Long.			89 Yards Long.			90 Yards Long.		
Yards.	*A.*	*R.*	*P.*	*A.*	*R.*	*P.*	*A.*	*R.*	*P.*	*A.*	*R.*	*P.*	*A.*	*R.*	*P.*
1	0	0	3	0	0	3	0	0	3	0	0	3	0	0	3
2	0	0	6	0	0	6	0	0	6	0	0	6	0	0	6
3	0	0	9	0	0	9	0	0	9	0	0	9	0	0	9
4	0	0	11	0	0	12	0	0	12	0	0	12	0	0	12
5	0	0	14	0	0	14	0	0	15	0	0	15	0	0	15
6	0	0	17	0	0	17	0	0	17	0	0	18	0	0	18
7	0	0	20	0	0	20	0	0	20	0	0	21	0	0	21
8	0	0	23	0	0	23	0	0	23	0	0	24	0	0	24
9	0	0	26	0	0	26	0	0	26	0	0	26	0	0	27
10	0	0	28	0	0	29	0	0	29	0	0	29	0	0	30
20	0	1	17	0	1	18	0	1	18	0	1	19	0	1	20
30	0	2	5	0	2	6	0	2	7	0	2	8	0	2	9
40	0	2	34	0	2	35	0	2	36	0	2	38	0	2	39
50	0	3	22	0	3	24	0	3	25	0	3	27	0	3	29
60	1	0	11	1	0	13	1	0	15	1	0	17	1	0	19
70	1	0	39	1	1	1	1	1	4	1	1	6	1	1	8
80	1	1	27	1	1	30	1	1	33	1	1	35	1	1	38
90	1	2	16	1	2	19	1	2	22	1	2	25	1	2	28
100	1	3	4	1	3	8	1	3	11	1	3	14	1	3	18
200	3	2	9	3	2	15	3	2	22	3	2	28	3	2	35
300	5	1	13	5	1	23	5	1	33	5	2	3	5	2	13
400	7	0	17	7	0	30	7	1	4	7	1	17	7	1	30
500	8	3	21	8	3	38	9	0	15	9	0	31	9	1	8

From 91 to 95 Yards Long.

Width.	91 Yards Long.			92 Yards Long.			93 Yards Long.			94 Yards Long.			95 Yards Long.		
Yards.	*A.*	*R.*	*P.*	*A.*	*R.*	*P.*	*A.*	*R.*	*P.*	*A.*	*R.*	*P.*	*A.*	*R.*	*P.*
1	0	0	3	0	0	3	0	0	3	0	0	3	0	0	3
2	0	0	6	0	0	6	0	0	6	0	0	6	0	0	6
3	0	0	9	0	0	9	0	0	9	0	0	9	0	0	9
4	0	0	12	0	0	12	0	0	12	0	0	12	0	0	13
5	0	0	15	0	0	15	0	0	15	0	0	16	0	0	16
6	0	0	18	0	0	18	0	0	18	0	0	19	0	0	19
7	0	0	21	0	0	21	0	0	22	0	0	22	0	0	22
8	0	0	24	0	0	24	0	0	25	0	0	25	0	0	25
9	0	0	27	0	0	27	0	0	28	0	0	28	0	0	28
10	0	0	30	0	0	30	0	0	31	0	0	31	0	0	31
20	0	1	20	0	1	21	0	1	21	0	1	22	0	1	23
30	0	2	10	0	2	11	0	2	12	0	2	13	0	2	14
40	0	3	0	0	3	2	0	3	3	0	3	4	0	3	6
50	0	3	30	0	3	32	0	3	34	0	3	35	0	3	37
60	1	0	20	1	0	22	1	0	24	1	0	26	1	0	28
70	1	1	11	1	1	13	1	1	15	1	1	18	1	1	20
80	1	2	1	1	2	3	1	2	6	1	2	9	1	2	11
90	1	2	31	1	2	34	1	2	37	1	3	0	1	3	3
100	1	3	21	1	3	24	1	3	27	1	3	31	1	3	34
200	3	3	2	3	3	8	3	3	15	3	3	21	3	3	28
300	5	2	22	5	2	32	5	3	2	5	3	12	5	3	22
400	7	2	3	7	2	17	7	2	30	7	3	3	7	3	16
500	9	1	24	9	2	1	9	2	17	9	2	34	9	3	10

From 96 to 100 Yards Long.

Width.	96 Yards Long.	97 Yards Long.	98 Yards Long.	99 Yards Long.	100 Yards Long.
Yards.	*A. R. P.*	*A. R. P.*	*A. R. P.*	*A. R. P.*	*A. R. P.*
1	0 0 3	0 0 3	0 0 3	0 0 3	0 0 3
2	0 0 6	0 0 6	0 0 6	0 0 7	0 0 7
3	0 0 10	0 0 10	0 0 10	0 0 10	0 0 10
4	0 0 13	0 0 13	0 0 13	0 0 13	0 0 13
5	0 0 16	0 0 16	0 0 16	0 0 16	0 0 17
6	0 0 19	0 0 19	0 0 19	0 0 20	0 0 20
7	0 0 22	0 0 22	0 0 23	0 0 23	0 0 23
8	0 0 25	0 0 26	0 0 26	0 0 26	0 0 26
9	0 0 29	0 0 29	0 0 29	0 0 29	0 0 30
10	0 0 32	0 0 32	0 0 32	0 0 33	0 0 33
20	0 1 23	0 1 24	0 1 25	0 1 25	0 1 26
30	0 2 15	0 2 16	0 2 17	0 2 18	0 2 19
40	0 3 7	0 3 8	0 3 10	0 3 11	0 3 12
50	0 3 39	1 0 0	1 0 2	1 0 4	1 0 5
60	1 0 30	1 0 32	1 0 34	1 0 36	1 0 38
70	1 1 22	1 1 24	1 1 27	1 1 29	1 1 31
80	1 2 14	1 2 17	1 2 19	1 2 22	1 2 24
90	1 3 6	1 3 9	1 3 12	1 3 15	1 3 18
100	1 3 37	2 0 1	2 0 4	2 0 7	2 0 11
200	3 3 35	4 0 1	4 0 8	4 0 15	4 0 21
300	5 3 32	6 0 2	6 0 12	6 0 22	6 0 32
400	7 3 29	8 0 3	8 0 16	8 0 29	8 1 2
500	9 3 27	10 0 3	10 0 20	10 0 36	10 1 13

From 101 to 105 Yards Long.

Width.	101 Yards Long.	102 Yards Long.	103 Yards Long.	104 Yards Long.	105 Yards Long.
Yards.	*A. R. P.*	*A. R. P.*	*A. R. P.*	*A. R. P.*	*A. R. P.*
1	0 0 3	0 0 3	0 0 3	0 0 3	0 0 3
2	0 0 7	0 0 7	0 0 7	0 0 7	0 0 7
3	0 0 10	0 0 10	0 0 10	0 0 10	0 0 10
4	0 0 13	0 0 13	0 0 14	0 0 14	0 0 14
5	0 0 17	0 0 17	0 0 17	0 0 17	0 0 17
6	0 0 20	0 0 20	0 0 20	0 0 21	0 0 21
7	0 0 23	0 0 24	0 0 24	0 0 24	0 0 24
8	0 0 27	0 0 27	0 0 27	0 0 28	0 0 28
9	0 0 30	0 0 30	0 0 31	0 0 31	0 0 31
10	0 0 33	0 0 34	0 0 34	0 0 34	0 0 35
20	0 1 27	0 1 27	0 1 28	0 1 29	0 1 29
30	0 2 20	0 2 21	0 2 22	0 2 23	0 2 24
40	0 3 14	0 3 15	0 3 16	0 3 18	0 3 19
50	1 0 7	1 0 9	1 0 10	1 0 12	1 0 14
60	1 1 0	1 1 2	1 1 4	1 1 6	1 1 8
70	1 1 34	1 1 36	1 1 38	1 2 1	1 2 3
80	1 2 27	1 2 30	1 2 32	1 2 35	1 2 38
90	1 3 20	1 3 23	1 3 26	1 3 29	1 3 32
100	2 0 14	2 0 17	2 0 20	2 0 24	2 0 27
200	4 0 28	4 0 34	4 1 1	4 1 8	4 1 14
300	6 1 2	6 1 12	6 1 21	6 1 31	6 2 1
400	8 1 16	8 1 29	8 2 2	8 2 15	8 2 28
500	10 1 29	10 2 6	10 2 22	10 2 39	10 3 16

From 106 to 110 Yards Long.

Width.	106 Yards Long.	107 Yards Long.	108 Yards Long.	109 Yards Long.	110 Yards Long.
Yards.	*A. R. P.*	*A. R. P.*	*A. R. P.*	*A. R. P.*	*A. R. P.*
1	0 0 4	0 0 4	0 0 4	0 0 4	0 0 4
2	0 0 7	0 0 7	0 0 7	0 0 7	0 0 7
3	0 0 11	0 0 11	0 0 11	0 0 11	0 0 11
4	0 0 14	0 0 14	0 0 14	0 0 14	0 0 15
5	0 0 18	0 0 18	0 0 18	0 0 18	0 0 18
6	0 0 21	0 0 21	0 0 21	0 0 22	0 0 22
7	0 0 25	0 0 25	0 0 25	0 0 25	0 0 25
8	0 0 28	0 0 28	0 0 29	0 0 29	0 0 29
9	0 0 32	0 0 32	0 0 32	0 0 32	0 0 33
10	0 0 35	0 0 35	0 0 36	0 0 36	0 0 36
20	0 1 30	0 1 31	0 1 31	0 1 32	0 1 33
30	0 2 25	0 2 26	0 2 27	0 2 28	0 2 29
40	0 3 20	0 3 21	0 3 23	0 3 24	0 3 25
50	1 0 15	1 0 17	1 0 19	1 0 20	1 0 22
60	1 1 10	1 1 12	1 1 14	1 1 16	1 1 18
70	1 2 5	1 2 8	1 2 10	1 2 12	1 2 15
80	1 3 0	1 3 3	1 3 6	1 3 8	1 3 11
90	1 3 35	1 3 38	2 0 1	2 0 4	2 0 7
100	2 0 30	2 0 34	2 0 37	2 1 0	2 1 4
200	4 1 21	4 1 27	4 1 34	4 2 1	4 2 7
300	6 2 11	6 2 21	6 2 31	6 3 1	6 3 11
400	8 3 2	8 3 15	8 3 28	9 0 1	9 0 15
500	10 3 32	11 0 9	11 0 25	11 1 2	11 1 18

From 111 to 115 Yards Long.

Width.	111 Yards Long.	112 Yards Long.	113 Yards Long.	114 Yards Long.	115 Yards Long.
Yards.	*A. R. P.*	*A. R. P.*	*A. R. P.*	*A. R. P.*	*A. R. P.*
1	0 0 4	0 0 4	0 0 4	0 0 4	0 0 4
2	0 0 7	0 0 7	0 0 7	0 0 8	0 0 8
3	0 0 11	0 0 11	0 0 11	0 0 11	0 0 11
4	0 0 15	0 0 15	0 0 15	0 0 15	0 0 15
5	0 0 18	0 0 19	0 0 19	0 0 19	0 0 19
6	0 0 22	0 0 22	0 0 22	0 0 23	0 0 23
7	0 0 26	0 0 26	0 0 26	0 0 26	0 0 27
8	0 0 29	0 0 30	0 0 30	0 0 30	0 0 30
9	0 0 33	0 0 33	0 0 34	0 0 34	0 0 34
10	0 0 37	0 0 37	0 0 37	0 0 38	0 0 38
20	0 1 33	0 1 34	0 1 35	0 1 35	0 1 36
30	0 2 30	0 2 31	0 2 32	0 2 33	0 2 34
40	0 3 27	0 3 28	0 3 29	0 3 31	0 3 32
50	1 0 23	1 0 25	1 0 27	1 0 28	1 0 30
60	1 1 20	1 1 22	1 1 24	1 1 26	1 1 28
70	1 2 17	1 2 19	1 2 21	1 2 24	1 2 26
80	1 3 14	1 3 16	1 3 19	1 3 21	1 3 24
90	2 0 10	2 0 13	2 0 16	2 0 19	2 0 22
100	2 1 7	2 1 10	2 1 14	2 1 17	2 1 20
200	4 2 14	4 2 20	4 2 27	4 2 34	4 3 0
300	6 3 21	6 3 31	7 0 1	7 0 11	7 0 20
400	9 0 28	9 1 1	9 1 14	9 1 27	9 2 1
500	11 1 35	11 2 11	11 2 28	11 3 4	11 3 21

From 116 to 120 Yards Long.

Width.	116 Yards Long.	117 Yards Long.	118 Yards Long.	119 Yards Long.	120 Yards Long.
Yards.	*A. R. P.*	*A. R. P.*	*A. R. P.*	*A. R. P.*	*A. R. P.*
1	0 0 4	0 0 4	0 0 4	0 0 4	0 0 4
2	0 0 8	0 0 8	0 0 8	0 0 8	0 0 8
3	0 0 12	0 0 12	0 0 12	0 0 12	0 0 12
4	0 0 15	0 0 15	0 0 16	0 0 16	0 0 16
5	0 0 19	0 0 19	0 0 20	0 0 20	0 0 20
6	0 0 23	0 0 23	0 0 23	0 0 24	0 0 24
7	0 0 27	0 0 27	0 0 27	0 0 28	0 0 28
8	0 0 31	0 0 31	0 0 31	0 0 31	0 0 32
9	0 0 35	0 0 35	0 0 35	0 0 35	0 0 36
10	0 0 38	0 0 39	0 0 39	0 0 39	0 1 0
20	0 1 37	0 1 37	0 1 38	0 1 39	0 1 39
30	0 2 35	0 2 36	0 2 37	0 2 38	0 2 39
40	0 3 33	0 3 35	0 3 36	0 3 37	0 3 39
50	1 0 32	1 0 33	1 0 35	1 0 37	1 0 38
60	1 1 30	1 1 32	1 1 34	1 1 36	1 1 38
70	1 2 28	1 2 31	1 2 33	1 2 35	1 2 38
80	1 3 27	1 3 29	1 3 32	1 3 35	1 3 37
90	2 0 25	2 0 28	2 0 31	2 0 34	2 0 37
100	2 1 23	2 1 27	2 1 30	2 1 33	2 1 37
200	4 3 7	4 3 14	4 3 20	4 3 27	4 3 33
300	7 0 30	7 1 0	7 1 10	7 1 20	7 1 30
400	9 2 14	9 2 27	9 3 0	9 3 14	9 3 27
500	11 3 37	12 0 14	12 0 30	12 1 7	12 1 23

From 121 to 125 Yards Long.

Width.	121 Yards Long.	122 Yards Long.	123 Yards Long.	124 Yards Long.	125 Yards Long.
Yards.	*A. R. P.*	*A. R. P.*	*A. R. P.*	*A. R. P.*	*A. R. P.*
1	0 0 4	0 0 4	0 0 4	0 0 4	0 0 4
2	0 0 8	0 0 8	0 0 8	0 0 8	0 0 8
3	0 0 12	0 0 12	0 0 12	0 0 12	0 0 12
4	0 0 16	0 0 16	0 0 16	0 0 16	0 0 17
5	0 0 20	0 0 20	0 0 20	0 0 20	0 0 21
6	0 0 24	0 0 24	0 0 24	0 0 25	0 0 25
7	0 0 28	0 0 28	0 0 28	0 0 29	0 0 29
8	0 0 32	0 0 32	0 0 33	0 0 33	0 0 33
9	0 0 36	0 0 36	0 0 37	0 0 37	0 0 37
10	0 1 0	0 1 0	0 1 1	0 1 1	0 1 1
20	0 2 0	0 2 1	0 2 1	0 2 2	0 2 3
30	0 3 0	0 3 1	0 3 2	0 3 3	0 3 4
40	1 0 0	1 0 1	1 0 3	1 0 4	1 0 5
50	1 1 0	1 1 2	1 1 3	1 1 5	1 1 7
60	1 2 0	1 2 2	1 2 4	1 2 6	1 2 8
70	1 3 0	1 3 2	1 3 5	1 3 7	1 3 9
80	2 0 0	2 0 3	2 0 5	2 0 8	2 0 11
90	2 1 0	2 1 3	2 1 6	2 1 9	2 1 12
100	2 2 0	2 2 3	2 2 7	2 2 10	2 2 13
200	5 0 0	5 0 7	5 0 13	5 0 20	5 0 26
300	7 2 0	7 2 10	7 2 20	7 2 30	7 3 0
400	10 0 0	10 0 13	10 0 26	10 1 0	10 1 13
500	12 2 0	12 2 17	12 2 33	12 3 10	12 3 26

From 126 to 130 Yards Long.

Width.	126 Yards Long.	127 Yards Long.	128 Yards Long.	129 Yards Long.	130 Yards Long.
Yards.	*A. R. P.*	*A. R. P.*	*A. R. P.*	*A. R. P.*	*A. R. P.*
1	0 0 4	0 0 4	0 0 4	0 0 4	0 0 4
2	0 0 8	0 0 8	0 0 8	0 0 9	0 0 9
3	0 0 12	0 0 13	0 0 13	0 0 13	0 0 13
4	0 0 17	0 0 17	0 0 17	0 0 17	0 0 17
5	0 0 21	0 0 21	0 0 21	0 0 21	0 0 21
6	0 0 25	0 0 25	0 0 25	0 0 26	0 0 26
7	0 0 29	0 0 29	0 0 30	0 0 30	0 0 30
8	0 0 33	0 0 34	0 0 34	0 0 34	0 0 34
9	0 0 37	0 0 38	0 0 38	0 0 38	0 0 39
10	0 1 2	0 1 2	0 1 2	0 1 3	0 1 3
20	0 2 3	0 2 4	0 2 5	0 2 5	0 2 6
30	0 3 5	0 3 6	0 3 7	0 3 8	0 3 9
40	1 0 7	1 0 8	1 0 9	1 0 11	1 0 12
50	1 1 8	1 1 10	1 1 12	1 1 13	1 1 15
60	1 2 10	1 2 12	1 2 14	1 2 16	1 2 18
70	1 3 12	1 3 14	1 3 16	1 3 19	1 3 21
80	2 0 13	2 0 16	2 0 19	2 0 21	2 0 24
90	2 1 15	2 1 18	2 1 21	2 1 24	2 1 27
100	2 2 17	2 2 20	2 2 23	2 2 26	2 2 30
200	5 0 33	5 1 0	5 1 6	5 1 13	5 1 20
300	7 3 10	7 3 20	7 3 29	7 3 39	8 0 9
400	10 1 26	10 1 39	10 2 13	10 2 26	10 2 39
500	13 0 3	13 0 19	13 0 36	13 1 12	13 1 29

From 131 to 135 Yards Long.

Width	131 Yards Long.	132 Yards Long.	133 Yards Long.	134 Yards Long.	135 Yards Long.
Yards.	A. R. P.	A. R. P.	A. R. P.	A. R. P.	A. R. P.
1	0 0 4	0 0 4	0 0 4	0 0 4	0 0 4
2	0 0 9	0 0 9	0 0 9	0 0 9	0 0 9
3	0 0 13	0 0 13	0 0 13	0 0 13	0 0 13
4	0 0 17	0 0 17	0 0 18	0 0 18	0 0 18
5	0 0 22	0 0 22	0 0 22	0 0 22	0 0 22
6	0 0 26	0 0 26	0 0 26	0 0 27	0 0 27
7	0 0 30	0 0 31	0 0 31	0 0 31	0 0 31
8	0 0 35	0 0 35	0 0 35	0 0 35	0 0 36
9	0 0 39	0 0 39	0 1 0	0 1 0	0 1 0
10	0 1 3	0 1 4	0 1 4	0 1 4	0 1 5
20	0 2 7	0 2 7	0 2 8	0 2 9	0 2 9
30	0 3 10	0 3 11	0 3 12	0 3 13	0 3 14
40	1 0 13	1 0 15	1 0 16	1 0 17	1 0 19
50	1 1 17	1 1 18	1 1 20	1 1 21	1 1 23
60	1 2 20	1 2 22	1 2 24	1 2 26	1 2 28
70	1 3 23	1 3 25	1 3 28	1 3 30	1 3 32
80	2 0 26	2 0 29	2 0 32	2 0 34	2 0 37
90	2 1 30	2 1 33	2 1 36	2 1 39	2 2 2
100	2 2 33	2 2 36	2 3 0	2 3 3	2 3 6
200	5 1 26	5 1 33	5 1 39	5 2 6	5 2 13
300	8 0 19	8 0 29	8 0 39	8 1 9	8 1 19
400	10 3 12	10 3 25	10 3 39	11 0 12	11 0 25
500	13 2 5	13 2 22	13 2 38	13 3 15	13 3 31

From 136 to 140 Yards Long.

Width.	136 Yards Long.	137 Yards Long.	138 Yards Long.	139 Yards Long.	140 Yards Long.
Yards.	*A. R. P.*	*A. R. P.*	*A. R. P.*	*A. R. P.*	*A. R. P.*
1	0 0 4	0 0 5	0 0 5	0 0 5	0 0 5
2	0 0 9	0 0 9	0 0 9	0 0 9	0 0 9
3	0 0 13	0 0 14	0 0 14	0 0 14	0 0 14
4	0 0 18	0 0 18	0 0 18	0 0 18	0 0 19
5	0 0 22	0 0 23	0 0 23	0 0 23	0 0 23
6	0 0 27	0 0 27	0 0 27	0 0 28	0 0 28
7	0 0 31	0 0 32	0 0 32	0 0 32	0 0 32
8	0 0 36	0 0 36	0 0 36	0 0 37	0 0 37
9	0 1 0	0 1 1	0 1 1	0 1 1	0 1 2
10	0 1 5	0 1 5	0 1 6	0 1 6	0 1 6
20	0 2 10	0 2 11	0 2 11	0 2 12	0 2 13
30	0 3 15	0 3 16	0 3 17	0 3 18	0 3 19
40	1 0 20	1 0 21	1 0 22	1 0 24	1 0 25
50	1 1 25	1 1 26	1 1 28	1 1 30	1 1 31
60	1 2 30	1 2 32	1 2 34	1 2 36	1 2 38
70	1 3 35	1 3 37	1 3 39	2 0 2	2 0 4
80	2 1 0	2 1 2	2 1 5	2 1 8	2 1 10
90	2 2 5	2 2 8	2 2 11	2 2 14	2 2 17
100	2 3 10	2 3 13	2 3 16	2 3 20	2 3 23
200	5 2 19	5 2 26	5 2 32	5 2 39	5 3 6
300	8 1 29	8 1 39	8 2 9	8 2 19	8 2 28
400	11 0 38	11 1 12	11 1 25	11 1 38	11 2 11
500	14 0 8	14 0 24	14 1 1	14 1 18	14 1 34

From 141 to 145 Yards Long.

Width.	141 Yards Long.	142 Yards Long.	143 Yards Long.	144 Yards Long.	145 Yards Long.
Yards.	*A. R. P.*	*A. R. P.*	*A. R. P.*	*A. R. P.*	*A. R. P.*
1	0 0 5	0 0 5	0 0 5	0 0 5	0 0 5
2	0 0 9	0 0 9	0 0 9	0 0 10	0 0 10
3	0 0 14	0 0 14	0 0 14	0 0 14	0 0 14
4	0 0 19	0 0 19	0 0 19	0 0 19	0 0 19
5	0 0 23	0 0 23	0 0 24	0 0 24	0 0 24
6	0 0 28	0 0 28	0 0 28	0 0 29	0 0 29
7	0 0 33	0 0 33	0 0 33	0 0 33	0 0 34
8	0 0 37	0 0 38	0 0 38	0 0 38	0 0 38
9	0 1 2	0 1 2	0 1 3	0 1 3	0 1 3
10	0 1 7	0 1 7	0 1 7	0 1 8	0 1 8
20	0 2 13	0 2 14	0 2 15	0 2 15	0 2 16
30	0 3 20	0 3 21	0 3 22	0 3 23	0 3 24
40	1 0 26	1 0 28	1 0 29	1 0 30	1 0 32
50	1 1 33	1 1 35	1 1 36	1 1 38	1 2 0
60	1 3 0	1 3 2	1 3 4	1 3 6	1 3 8
70	2 0 6	2 0 9	2 0 11	2 0 13	2 0 16
80	2 1 13	2 1 16	2 1 18	2 1 21	2 1 23
90	2 2 20	2 2 22	2 2 25	2 2 28	2 2 31
100	2 3 26	2 3 29	2 3 33	2 3 36	2 3 39
200	5 3 12	5 3 19	5 3 25	5 3 32	5 3 39
300	8 2 38	8 3 8	8 3 18	8 3 28	8 3 38
400	11 2 24	11 2 38	11 3 11	11 3 24	11 3 37
500	14 2 11	14 2 27	14 3 4	14 3 20	14 3 37

From 146 to 150 Yards Long.

Width.	146 Yards Long.	147 Yards Long.	148 Yards Long.	149 Yards Long.	150 Yards Long.
Yards.	*A. R. P.*	*A. R. P.*	*A. R. P.*	*A. R. P.*	*A. R. P.*
1	0 0 5	0 0 5	0 0 5	0 0 5	0 0 5
2	0 0 10	0 0 10	0 0 10	0 0 10	0 0 10
3	0 0 14	0 0 15	0 0 15	0 0 15	0 0 15
4	0 0 19	0 0 19	0 0 20	0 0 20	0 0 20
5	0 0 24	0 0 24	0 0 24	0 0 25	0 0 25
6	0 0 29	0 0 29	0 0 29	0 0 30	0 0 30
7	0 0 34	0 0 34	0 0 34	0 0 34	0 0 35
8	0 0 39	0 0 39	0 0 39	0 0 39	0 1 0
9	0 1 3	0 1 4	0 1 4	0 1 4	0 1 5
10	0 1 8	0 1 9	0 1 9	0 1 9	0 1 10
20	0 2 17	0 2 17	0 2 18	0 2 19	0 2 19
30	0 3 25	0 3 26	0 3 27	0 3 28	0 3 29
40	1 0 33	1 0 34	1 0 36	1 0 37	1 0 38
50	1 2 1	1 2 3	1 2 5	1 2 6	1 2 8
60	1 3 10	1 3 12	1 3 14	1 3 16	1 3 18
70	2 0 18	2 0 20	2 0 22	2 0 25	2 0 27
80	2 1 26	2 1 29	2 1 31	2 1 34	2 1 37
90	2 2 34	2 2 37	2 3 0	2 3 3	2 3 6
100	3 0 3	3 0 6	3 0 9	3 0 13	3 0 16
200	6 0 5	6 0 12	6 0 19	6 0 25	6 0 32
300	9 0 8	9 0 18	9 0 28	9 0 38	9 1 8
400	12 0 11	12 0 24	12 0 37	12 1 10	12 1 23
500	15 0 13	15 0 30	15 1 6	15 1 23	15 1 39

From 151 to 155 Yards Long.

Width.	151 Yards Long.	152 Yards Long.	153 Yards Long.	154 Yards Long.	155 Yards Long.
Yards.	*A. R. P.*	*A. R. P.*	*A. R. P.*	*A. R. P.*	*A. R. P.*
1	0 0 5	0 0 5	0 0 5	0 0 5	0 0 5
2	0 0 10	0 0 10	0 0 10	0 0 10	0 0 10
3	0 0 15	0 0 15	0 0 15	0 0 15	0 0 15
4	0 0 20	0 0 20	0 0 20	0 0 20	0 0 20
5	0 0 25	0 0 25	0 0 25	0 0 25	0 0 26
6	0 0 30	0 0 30	0 0 30	0 0 31	0 0 31
7	0 0 35	0 0 35	0 0 35	0 0 36	0 0 36
8	0 1 0	0 1 0	0 1 0	0 1 1	0 1 1
9	0 1 5	0 1 5	0 1 6	0 1 6	0 1 6
10	0 1 10	0 1 10	0 1 11	0 1 11	0 1 11
20	0 2 20	0 2 20	0 2 21	0 2 22	0 2 22
30	0 3 30	0 3 31	0 3 32	0 3 33	0 3 34
40	1 1 0	1 1 1	1 1 2	1 1 4	1 1 5
50	1 2 10	1 2 11	1 2 13	1 2 15	1 2 16
60	1 3 20	1 3 21	1 3 23	1 3 25	1 3 27
70	2 0 29	2 0 32	2 0 34	2 0 36	2 0 39
80	2 1 39	2 2 2	2 2 5	2 2 7	2 2 10
90	2 3 9	2 3 12	2 3 15	2 3 18	2 3 21
100	3 0 19	3 0 22	3 0 26	3 0 29	3 0 32
200	6 0 38	6 1 5	6 1 12	6 1 18	6 1 25
300	9 1 18	9 1 27	9 1 37	9 2 7	9 2 17
400	12 1 37	12 2 10	12 2 23	12 2 36	12 3 10
500	15 2 16	15 2 32	15 3 9	15 3 25	16 0 2

From 156 to 160 Yards Long.

Width.	156 Yards Long.	157 Yards Long.	158 Yards Long.	159 Yards Long.	160 Yards Long.
Yards.	*A. R. P.*	*A. R. P.*	*A. R. P.*	*A. R. P.*	*A. R. P.*
1	0 0 5	0 0 5	0 0 5	0 0 5	0 0 5
2	0 0 10	0 0 10	0 0 10	0 0 11	0 0 11
3	0 0 15	0 0 16	0 0 16	0 0 16	0 0 16
4	0 0 21	0 0 21	0 0 21	0 0 21	0 0 21
5	0 0 26	0 0 26	0 0 26	0 0 26	0 0 26
6	0 0 31	0 0 31	0 0 31	0 0 32	0 0 32
7	0 0 36	0 0 36	0 0 37	0 0 37	0 0 37
8	0 1 1	0 1 2	0 1 2	0 1 2	0 1 2
9	0 1 6	0 1 7	0 1 7	0 1 7	0 1 8
10	0 1 12	0 1 12	0 1 12	0 1 13	0 1 13
20	0 2 23	0 2 24	0 2 24	0 2 25	0 2 26
30	0 3 35	0 3 36	0 3 37	0 3 38	0 3 39
40	1 1 6	1 1 8	1 1 9	1 1 10	1 1 12
50	1 2 18	1 2 20	1 2 21	1 2 23	1 2 24
60	1 3 29	1 3 31	1 3 33	1 3 35	1 3 37
70	2 1 1	2 1 3	2 1 6	2 1 8	2 1 10
80	2 2 13	2 2 15	2 2 18	2 2 20	2 2 23
90	2 3 24	2 3 27	2 3 30	2 3 33	2 3 36
100	3 0 36	3 0 39	3 1 2	3 1 6	3 1 9
200	6 1 31	6 1 38	6 2 5	6 2 11	6 2 18
300	9 2 27	9 2 37	9 3 7	9 3 17	9 3 27
400	12 3 23	12 3 36	13 0 9	13 0 22	13 0 36
500	16 0 19	16 0 35	16 1 12	16 1 28	16 2 5

From 161 to 165 Yards Long.

Width.	161 Yards Long.	162 Yards Long.	163 Yards Long.	164 Yards Long.	165 Yards Long.
Yards.	*A. R. P.*	*A. R. P.*	*A. R. P.*	*A. R. P.*	*A. R. P.*
1	0 0 5	0 0 5	0 0 5	0 0 5	0 0 5
2	0 0 11	0 0 11	0 0 11	0 0 11	0 0 11
3	0 0 16	0 0 16	0 0 16	0 0 16	0 0 16
4	0 0 21	0 0 21	0 0 22	0 0 22	0 0 22
5	0 0 27	0 0 27	0 0 27	0 0 27	0 0 27
6	0 0 32	0 0 32	0 0 32	0 0 33	0 0 33
7	0 0 37	0 0 37	0 0 38	0 0 38	0 0 38
8	0 1 3	0 1 3	0 1 3	0 1 3	0 1 4
9	0 1 8	0 1 8	0 1 8	0 1 9	0 1 9
10	0 1 13	0 1 14	0 1 14	0 1 14	0 1 15
20	0 2 26	0 2 27	0 2 28	0 2 28	0 2 29
30	1 0 0	1 0 1	1 0 2	1 0 3	1 0 4
40	1 1 13	1 1 14	1 1 16	1 1 17	1 1 18
50	1 2 26	1 2 28	1 2 29	1 2 31	1 2 33
60	1 3 39	2 0 1	2 0 3	2 0 5	2 0 7
70	2 1 13	2 1 15	2 1 17	2 1 20	2 1 22
80	2 2 26	2 2 28	2 2 31	2 2 34	2 2 36
90	2 3 39	3 0 2	3 0 5	3 0 8	3 0 11
100	3 1 12	3 1 16	3 1 19	3 1 22	3 1 25
200	6 2 24	6 2 31	6 2 38	6 3 4	6 3 11
300	9 3 37	10 0 7	10 0 17	10 0 26	10 0 36
400	13 1 9	13 1 22	13 1 35	13 2 9	13 2 22
500	16 2 21	16 2 38	16 3 14	16 3 31	17 0 7

From 166 to 170 Yards Long.

Width.	166 Yards Long.	167 Yards Long.	168 Yards Long.	169 Yards Long.	170 Yards Long.
Yards.	*A. R. P.*	*A. R. P.*	*A. R. P.*	*A. R. P.*	*A. R. P.*
1	0 0 5	0 0 6	0 0 6	0 0 6	0 0 6
2	0 0 11	0 0 11	0 0 11	0 0 11	0 0 11
3	0 0 16	0 0 17	0 0 17	0 0 17	0 0 17
4	0 0 22	0 0 22	0 0 22	0 0 22	0 0 22
5	0 0 27	0 0 28	0 0 28	0 0 28	0 0 28
6	0 0 33	0 0 33	0 0 33	0 0 34	0 0 34
7	0 0 38	0 0 39	0 0 39	0 0 39	0 0 39
8	0 1 4	0 1 4	0 1 4	0 1 5	0 1 5
9	0 1 9	0 1 10	0 1 10	0 1 10	0 1 11
10	0 1 15	0 1 15	0 1 16	0 1 16	0 1 16
20	0 2 30	0 2 30	0 2 31	0 2 32	0 2 32
30	1 0 5	1 0 6	1 0 7	1 0 8	1 0 9
40	1 1 20	1 1 21	1 1 22	1 1 23	1 1 25
50	1 2 34	1 2 36	1 2 38	1 2 39	1 3 1
60	2 0 9	2 0 11	2 0 13	2 0 15	2 0 17
70	2 1 24	2 1 26	2 1 29	2 1 31	2 1 33
80	2 2 39	2 3 2	2 3 4	2 3 7	2 3 10
90	3 0 14	3 0 17	3 0 20	3 0 23	3 0 26
100	3 1 29	3 1 32	3 1 35	3 1 39	3 2 2
200	6 3 18	6 3 24	6 3 31	6 3 37	7 0 4
300	10 1 6	10 1 16	10 1 26	10 1 36	10 2 6
400	13 2 35	13 3 8	13 3 21	13 3 35	14 0 8
500	17 0 24	17 1 0	17 1 17	17 1 33	17 2 10

From 171 to 175 Yards Long.

Width.	171 Yards Long.	172 Yards Long.	173 Yards Long.	174 Yards Long.	175 Yards Long.
Yards.	*A. R. P.*	*A. R. P.*	*A. R. P.*	*A. R. P.*	*A. R. P.*
1	0 0 6	0 0 6	0 0 6	0 0 6	0 0 6
2	0 0 11	0 0 11	0 0 11	0 0 12	0 0 12
3	0 0 17	0 0 17	0 0 17	0 0 17	0 0 17
4	0 0 23	0 0 23	0 0 23	0 0 23	0 0 23
5	0 0 28	0 0 28	0 0 29	0 0 29	0 0 29
6	0 0 34	0 0 34	0 0 34	0 0 35	0 0 35
7	0 1 0	0 1 0	0 1 0	0 1 0	0 1 0
8	0 1 5	0 1 5	0 1 6	0 1 6	0 1 6
9	0 1 11	0 1 11	0 1 11	0 1 12	0 1 12
10	0 1 17	0 1 17	0 1 17	0 1 18	0 1 18
20	0 2 33	0 2 34	0 2 34	0 2 35	0 2 36
30	1 0 10	1 0 11	1 0 12	1 0 13	1 0 14
40	1 1 26	1 1 27	1 1 29	1 1 30	1 1 31
50	1 3 3	1 3 4	1 3 6	1 3 8	1 3 9
60	2 0 19	2 0 21	2 0 23	2 0 25	2 0 27
70	2 1 36	2 1 38	2 2 0	2 2 3	2 2 5
80	2 3 12	2 3 15	2 3 18	2 3 20	2 3 23
90	3 0 29	3 0 32	3 0 35	3 0 38	3 1 1
100	3 2 5	3 2 9	3 2 12	3 2 15	3 2 19
200	7 0 11	7 0 17	7 0 24	7 0 30	7 0 37
300	10 2 16	10 2 26	10 2 36	10 3 6	10 3 16
400	14 0 21	14 0 34	14 1 8	14 1 21	14 1 34
500	17 2 26	17 3 3	17 3 20	17 3 36	18 0 13

From 176 to 180 Yards Long.

Width.	176 Yards Long.	177 Yards Long.	178 Yards Long.	179 Yards Long.	180 Yards Long.
Yards.	*A. R. P.*	*A. R. P.*	*A. R. P.*	*A. R. P.*	*A. R. P.*
1	0 0 6	0 0 6	0 0 6	0 0 6	0 0 6
2	0 0 12	0 0 12	0 0 12	0 0 12	0 0 12
3	0 0 17	0 0 18	0 0 18	0 0 18	0 0 18
4	0 0 23	0 0 23	0 0 24	0 0 24	0 0 24
5	0 0 29	0 0 29	0 0 29	0 0 30	0 0 30
6	0 0 35	0 0 35	0 0 35	0 0 36	0 0 36
7	0 1 1	0 1 1	0 1 1	0 1 1	0 1 2
8	0 1 7	0 1 7	0 1 7	0 1 7	0 1 8
9	0 1 12	0 1 13	0 1 13	0 1 13	0 1 14
10	0 1 18	0 1 19	0 1 19	0 1 19	0 1 20
20	0 2 36	0 2 37	0 2 38	0 2 38	0 2 39
30	1 0 15	1 0 16	1 0 17	1 0 18	1 0 19
40	1 1 33	1 1 34	1 1 35	1 1 37	1 1 38
50	1 3 11	1 3 13	1 3 14	1 3 16	1 3 18
60	2 0 29	2 0 31	2 0 33	2 0 35	2 0 37
70	2 2 7	2 2 10	2 2 12	2 2 14	2 2 17
80	2 3 25	2 3 28	2 3 31	2 3 33	2 3 36
90	3 1 4	3 1 7	3 1 10	3 1 13	3 1 16
100	3 2 22	3 2 25	3 2 28	3 2 32	3 2 35
200	7 1 4	7 1 10	7 1 17	7 1 23	7 1 30
300	10 3 25	10 3 35	11 0 5	11 0 15	11 0 25
400	14 2 7	14 2 20	14 2 34	14 3 7	14 3 20
500	18 0 29	18 1 6	18 1 22	18 1 39	18 2 15

From 181 to 185 Yards Long.

Width.	181 Yards Long.	182 Yards Long.	183 Yards Long.	184 Yards Long.	185 Yards Long.
Yards.	*A. R. P.*	*A. R. P.*	*A. R. P.*	*A. R. P.*	*A. R. P.*
1	0 0 6	0 0 6	0 0 6	0 0 6	0 0 6
2	0 0 12	0 0 12	0 0 12	0 0 12	0 0 12
3	0 0 18	0 0 18	0 0 18	0 0 18	0 0 18
4	0 0 24	0 0 24	0 0 24	0 0 24	0 0 24
5	0 0 30	0 0 30	0 0 30	0 0 30	0 0 31
6	0 0 36	0 0 36	0 0 36	0 0 36	0 0 37
7	0 1 2	0 1 2	0 1 2	0 1 3	0 1 3
8	0 1 8	0 1 8	0 1 8	0 1 9	0 1 9
9	0 1 14	0 1 14	0 1 14	0 1 15	0 1 15
10	0 1 20	0 1 20	0 1 20	0 1 21	0 1 21
20	0 3 0	0 3 0	0 3 1	0 3 2	0 3 2
30	1 0 20	1 0 20	1 0 21	1 0 22	1 0 23
40	1 1 39	1 2 1	1 2 2	1 2 3	1 2 5
50	1 3 19	1 3 21	1 3 22	1 3 24	1 3 26
60	2 0 39	2 1 1	2 1 3	2 1 5	2 1 7
70	2 2 19	2 2 21	2 2 23	2 2 26	2 2 28
80	2 3 39	3 0 1	3 0 4	3 0 7	3 0 9
90	3 1 19	3 1 21	3 1 24	3 1 27	3 1 30
100	3 2 38	3 3 2	3 2 5	3 3 8	3 3 12
200	7 1 37	7 2 3	7 2 10	7 2 17	7 2 23
300	11 0 35	11 1 5	11 1 15	11 1 25	11 1 35
400	14 3 33	15 0 7	15 0 20	15 0 33	15 1 6
500	18 2 32	18 3 8	18 3 25	19 0 1	19 0 18

From 186 to 190 Yards Long.

Width.	186 Yards Long.	187 Yards Long.	188 Yards Long.	189 Yards Long.	190 Yards Long.
Yards.	*A. R. P.*	*A. R. P.*	*A. R. P.*	*A. R. P.*	*A. R. P.*
1	0 0 6	0 0 6	0 0 6	0 0 6	0 0 6
2	0 0 12	0 0 12	0 0 12	0 0 12	0 0 13
3	0 0 18	0 0 19	0 0 19	0 0 19	0 0 19
4	0 0 25	0 0 25	0 0 25	0 0 25	0 0 25
5	0 0 31	0 0 31	0 0 31	0 0 31	0 0 31
6	0 0 37	0 0 37	0 0 37	0 0 37	0 0 38
7	0 1 3	0 1 3	0 1 4	0 1 4	0 1 4
8	0 1 9	0 1 9	0 1 10	0 1 10	0 1 10
9	0 1 15	0 1 16	0 1 16	0 1 16	0 1 17
10	0 1 21	0 1 22	0 1 22	0 1 22	0 1 23
20	0 3 3	0 3 4	0 3 4	0 3 5	0 3 6
30	1 0 24	1 0 25	1 0 26	1 0 27	1 0 28
40	1 2 6	1 2 7	1 2 9	1 2 10	1 2 11
50	1 3 27	1 3 29	1 3 31	1 3 32	1 3 34
60	2 1 9	2 1 11	2 1 13	2 1 15	2 1 17
70	2 2 30	2 2 33	2 2 35	2 2 37	2 3 0
80	3 0 12	3 0 15	3 0 17	3 0 20	3 0 22
90	3 1 33	3 1 36	3 1 39	3 2 2	3 2 5
100	3 3 15	3 3 18	3 3 21	3 3 25	3 3 28
200	7 2 30	7 2 36	7 3 3	7 3 10	7 3 16
300	11 2 5	11 2 15	11 2 24	11 2 34	11 3 4
400	15 1 20	15 1 33	15 2 6	15 2 19	15 2 32
500	19 0 34	19 1 11	19 1 27	19 2 4	19 2 20

From 191 to 195 Yards Long.

Width.	191 Yards Long.	192 Yards Long.	193 Yards Long.	194 Yards Long.	195 Yards Long.
Yards.	*A. R. P.*	*A. R. P.*	*A. R. P.*	*A. R. P.*	*A. R. P.*
1	0 0 6	0 0 6	0 0 6	0 0 6	0 0 6
2	0 0 13	0 0 13	0 0 13	0 0 13	0 0 13
3	0 0 19	0 0 19	0 0 19	0 0 19	0 0 19
4	0 0 25	0 0 25	0 0 26	0 0 26	0 0 26
5	0 0 32	0 0 32	0 0 32	0 0 32	0 0 32
6	0 0 38	0 0 38	0 0 38	0 0 38	0 0 39
7	0 1 4	0 1 4	0 1 5	0 1 5	0 1 5
8	0 1 11	0 1 11	0 1 11	0 1 11	0 1 12
9	0 1 17	0 1 17	0 1 17	0 1 18	0 1 18
10	0 1 23	0 1 23	0 1 24	0 1 24	0 1 24
20	0 3 6	0 3 7	0 3 8	0 3 8	0 3 9
30	1 0 29	1 0 30	1 0 31	1 0 32	1 0 33
40	1 2 13	1 2 14	1 2 15	1 2 17	1 2 18
50	1 3 36	1 3 37	1 3 39	2 0 1	2 0 2
60	2 1 19	2 1 21	2 1 23	2 1 25	2 1 27
70	2 3 2	2 3 4	2 3 7	2 3 9	2 3 11
80	3 0 25	3 0 28	3 0 30	3 0 33	3 0 36
90	3 2 8	3 2 11	3 2 14	3 2 17	3 2 20
100	3 3 31	3 3 35	3 3 38	4 0 1	4 0 5
200	7 3 23	7 3 29	7 3 36	8 0 3	8 0 9
300	11 3 14	11 3 24	11 3 34	12 0 4	12 0 14
400	15 3 6	15 3 19	15 3 32	16 0 5	16 0 19
500	19 2 37	19 3 14	19 3 30	20 0 7	20 0 23

From 196 to 200 Yards Long.

Width.	196 Yards Long.	197 Yards Long.	198 Yards Long.	199 Yards Long.	200 Yards Long.
Yards.	*A. R. P.*	*A. R. P.*	*A. R. P.*	*A. R. P.*	*A. R. P.*
1	0 0 6	0 0 7	0 0 7	0 0 7	0 0 7
2	0 0 13	0 0 13	0 0 13	0 0 13	0 0 13
3	0 0 19	0 0 20	0 0 20	0 0 20	0 0 20
4	0 0 26	0 0 26	0 0 26	0 0 26	0 0 26
5	0 0 32	0 0 33	0 0 33	0 0 33	0 0 33
6	0 0 39	0 0 39	0 0 39	0 0 39	0 1 0
7	0 1 5	0 1 6	0 1 6	0 1 6	0 1 6
8	0 1 12	0 1 12	0 1 12	0 1 13	0 1 13
9	0 1 18	0 1 19	0 1 19	0 1 19	0 1 20
10	0 1 25	0 1 25	0 1 25	0 1 26	0 1 26
20	0 3 10	0 3 10	0 3 11	0 3 12	0 3 12
30	1 0 34	1 0 35	1 0 36	1 0 37	1 0 38
40	1 2 19	1 2 20	1 2 22	1 2 23	1 2 24
50	2 0 4	2 0 6	2 0 7	2 0 9	2 0 11
60	2 1 29	2 1 31	2 1 33	2 1 35	2 1 37
70	2 3 14	2 3 16	2 3 18	2 3 20	2 3 23
80	3 0 38	3 1 1	3 1 4	3 1 6	3 1 9
90	3 2 23	3 2 26	3 2 29	3 2 32	3 2 35
100	4 0 8	4 0 11	4 0 15	4 0 18	4 0 21
200	8 0 16	8 0 22	8 0 29	8 0 36	8 1 2
300	12 0 24	12 0 34	12 1 4	12 1 14	12 1 23
400	16 0 32	16 1 5	16 1 18	16 1 31	16 2 5
500	20 1 0	20 1 16	20 1 33	20 2 9	20 2 26

From 201 to 205 Yards Long.

Width.	201 Yards Long.			202 Yards Long.			203 Yards Long.			204 Yards Long.			205 Yards Long.		
Yards.	*A.*	*R.*	*P.*	*A.*	*R.*	*P.*	*A.*	*R.*	*P.*	*A.*	*R.*	*P.*	*A.*	*R.*	*P.*
1	0	0	7	0	0	7	0	0	7	0	0	7	0	0	7
2	0	0	13	0	0	13	0	0	13	0	0	13	0	0	14
3	0	0	20	0	0	20	0	0	20	0	0	20	0	0	20
4	0	0	27	0	0	27	0	0	27	0	0	27	0	0	27
5	0	0	33	0	0	33	0	0	34	0	0	34	0	0	34
6	0	1	0	0	1	0	0	1	0	0	1	0	0	1	1
7	0	1	7	0	1	7	0	1	7	0	1	7	0	1	7
8	0	1	13	0	1	13	0	1	14	0	1	14	0	1	14
9	0	1	20	0	1	20	0	1	20	0	1	21	0	1	21
10	0	1	26	0	1	27	0	1	27	0	1	27	0	1	28
20	0	3	13	0	3	14	0	3	14	0	3	15	0	3	16
30	1	0	39	1	1	0	1	1	1	1	1	2	1	1	3
40	1	2	26	1	2	27	1	2	28	1	2	30	1	2	31
50	2	0	12	2	0	14	2	0	16	2	0	17	2	0	19
60	2	1	39	2	2	1	2	2	3	2	2	5	2	2	7
70	2	3	25	2	3	27	2	3	30	2	3	32	2	3	34
80	3	1	12	3	1	14	3	1	17	3	1	20	3	1	22
90	3	2	38	3	3	1	3	3	4	3	3	7	3	3	10
100	4	0	24	4	0	28	4	0	31	4	0	34	4	0	38
200	8	1	9	8	1	16	8	1	22	8	1	29	8	1	35
300	12	1	33	12	2	3	12	2	13	12	2	23	12	2	33
400	16	2	18	16	2	31	16	3	4	16	3	18	16	3	31
500	20	3	2	20	3	19	20	3	35	21	0	12	21	0	28

From 206 to 210 Yards Long.

Width.	206 Yards Long.	207 Yards Long.	208 Yards Long.	209 Yards Long.	210 Yards Long.
Yards.	*A. R. P.*	*A. R. P.*	*A. R. P.*	*A. R. P.*	*A. R. P.*
1	0 0 7	0 0 7	0 0 7	0 0 7	0 0 7
2	0 0 14	0 0 14	0 0 14	0 0 14	0 0 14
3	0 0 20	0 0 21	0 0 21	0 0 21	0 0 21
4	0 0 27	0 0 27	0 0 28	0 0 28	0 0 28
5	0 0 34	0 0 34	0 0 34	0 0 35	0 0 35
6	0 1 1	0 1 1	0 1 1	0 1 1	0 1 2
7	0 1 8	0 1 8	0 1 8	0 1 8	0 1 9
8	0 1 14	0 1 15	0 1 15	0 1 15	0 1 16
9	0 1 21	0 1 22	0 1 22	0 1 22	0 1 22
10	0 1 28	0 1 28	0 1 29	0 1 29	0 1 29
20	0 3 16	0 3 17	0 3 18	0 3 18	0 3 19
30	1 1 4	1 1 5	1 1 6	1 1 7	1 1 8
40	1 2 32	1 2 34	1 2 35	1 2 36	1 2 38
50	2 0 20	2 0 22	2 0 24	2 0 25	2 0 27
60	2 2 9	2 2 11	2 2 13	2 2 15	2 2 17
70	2 3 37	2 3 39	3 0 1	3 0 4	3 0 6
80	3 1 25	3 1 27	3 1 30	3 1 33	3 1 35
90	3 3 13	3 3 16	3 3 9	3 3 22	3 3 25
100	4 1 1	4 1 4	4 1 8	4 1 11	4 1 14
200	8 2 2	8 2 9	8 2 15	8 2 22	8 2 28
300	12 3 3	12 3 13	12 3 23	12 3 33	13 0 3
400	17 0 4	17 0 17	17 0 30	17 1 4	17 1 17
500	21 1 5	21 1 21	21 1 38	21 2 15	21 2 31

From 211 to 215 Yards Long.

Width.	211 Yards Long.	212 Yards Long.	213 Yards Long.	214 Yards Long.	215 Yards Long.
Yards	A. R. P.	A. R. P.	A. R. P.	A. R. P.	A. R. P.
1	0 0 7	0 0 7	0 0 7	0 0 7	0 0 7
2	0 0 14	0 0 14	0 0 14	0 0 14	0 0 14
3	0 0 21	0 0 21	0 0 21	0 0 21	0 0 21
4	0 0 28	0 0 28	0 0 28	0 0 28	0 0 28
5	0 0 35	0 0 35	0 0 35	0 0 35	0 0 36
6	0 1 2	0 1 2	0 1 2	0 1 2	0 1 3
7	0 1 9	0 1 9	0 1 9	0 1 10	0 1 10
8	0 1 16	0 1 16	0 1 16	0 1 17	0 1 17
9	0 1 23	0 1 23	0 1 23	0 1 24	0 1 24
10	0 1 30	0 1 30	0 1 30	0 1 31	0 1 31
20	0 3 20	0 3 20	0 3 21	0 3 21	0 3 22
30	1 1 9	1 1 10	1 1 11	1 1 12	1 1 13
40	1 2 39	1 3 0	1 3 2	1 3 3	1 3 4
50	2 0 29	2 0 30	2 0 32	2 0 34	2 0 35
60	2 2 19	2 2 20	2 2 22	2 2 24	2 2 26
70	3 0 8	3 0 11	3 0 13	3 0 15	3 0 18
80	3 1 38	3 2 1	3 2 3	3 2 6	3 2 9
90	3 3 28	3 3 31	3 3 34	3 3 37	4 0 0
100	4 1 18	4 1 21	4 1 24	4 1 27	4 1 31
200	8 2 35	8 3 2	8 3 8	8 3 15	8 3 21
300	13 0 13	13 0 22	13 0 32	13 1 2	13 1 12
400	17 1 30	17 2 3	17 2 17	17 2 30	17 3 3
500	21 3 8	21 3 24	22 0 1	22 0 17	22 0 34

From 216 to 220 Yards Long.

Width.	216 Yards Long.	217 Yards Long.	218 Yards Long.	219 Yards Long.	220 Yards Long.
Yards.	*A. R. P.*	*A. R. P.*	*A. R. P.*	*A. R. P.*	*A. R. P.*
1	0 0 7	0 0 7	0 0 7	0 0 7	0 0 7
2	0 0 14	0 0 14	0 0 14	0 0 14	0 0 15
3	0 0 21	0 0 22	0 0 22	0 0 22	0 0 22
4	0 0 29	0 0 29	0 0 29	0 0 29	0 0 29
5	0 0 36	0 0 36	0 0 36	0 0 36	0 0 36
6	0 1 3	0 1 3	0 1 3	0 1 3	0 1 4
7	0 1 10	0 1 10	0 1 10	0 1 11	0 1 11
8	0 1 17	0 1 17	0 1 18	0 1 18	0 1 18
9	0 1 24	0 1 25	0 1 25	0 1 25	0 1 25
10	0 1 31	0 1 32	0 1 32	0 1 32	0 1 33
20	0 3 23	0 3 23	0 3 24	0 3 25	0 3 25
30	1 1 14	1 1 15	1 1 16	1 1 17	1 1 18
40	1 3 6	1 3 7	1 3 8	1 3 10	1 3 11
50	2 0 37	2 0 39	2 1 0	2 1 2	2 1 4
60	2 2 28	2 2 30	2 2 32	2 2 34	2 2 36
70	3 0 20	3 0 22	3 0 24	3 0 27	3 0 29
80	3 2 11	3 2 14	3 2 17	3 2 19	3 2 22
90	4 0 3	4 0 6	4 0 9	4 0 12	4 0 15
100	4 1 34	4 1 37	4 2 1	4 2 4	4 2 7
200	8 3 28	8 3 35	9 0 1	9 0 8	9 0 15
300	13 1 22	13 1 32	13 2 2	13 2 12	13 2 22
400	17 3 16	17 3 29	18 0 3	18 0 16	18 0 29
500	22 1 10	22 1 27	22 2 3	22 2 20	22 2 36

[illegible] 221 to 225 Yards Long.

Width.	221 Yards Long.	222 Yards Long.	223 Yards Long.	224 Yards Long.	225 Yards Long.
Yards.	*A. R. P.*	*A. R. P.*	*A. R. P.*	*A. R. P.*	*A. R. P.*
1	0 0 7	0 0 7	0 0 7	0 0 7	0 0 7
2	0 0 15	0 0 15	0 0 15	0 0 15	0 0 15
3	0 0 22	0 0 22	0 0 22	0 0 22	0 0 22
4	0 0 29	0 0 29	0 0 29	0 0 30	0 0 30
5	0 0 37	0 0 37	0 0 37	0 0 37	0 0 37
6	0 1 4	0 1 4	0 1 4	0 1 4	0 1 5
7	0 1 11	0 1 11	0 1 12	0 1 12	0 1 12
8	0 1 18	0 1 19	0 1 19	0 1 19	0 1 20
9	0 1 26	0 1 26	0 1 26	0 1 27	0 1 27
10	0 1 33	0 1 33	0 1 34	0 1 34	0 1 34
20	0 3 26	0 3 27	0 3 27	0 3 28	0 3 29
30	1 1 19	1 1 20	1 1 21	1 1 22	1 1 23
40	1 3 12	1 3 14	1 3 15	1 3 16	1 3 18
50	2 1 5	2 1 7	2 1 9	2 1 10	2 1 12
60	2 2 38	2 3 0	2 3 2	2 3 4	2 3 6
70	3 0 31	3 0 34	3 0 36	3 0 38	3 1 1
80	3 2 24	3 2 27	3 2 30	3 2 32	3 2 35
90	4 0 18	4 0 20	4 0 23	4 0 26	4 0 29
100	4 2 11	4 2 14	4 2 17	4 2 20	4 2 24
200	9 0 21	9 0 28	9 0 34	9 1 1	9 1 8
300	13 2 32	13 3 2	13 3 12	13 3 21	13 3 31
400	18 1 2	18 1 16	18 1 29	18 2 2	18 2 15
500	22 3 13	22 3 29	23 0 6	23 0 22	23 0 39

From 226 to 230 Yards Long.

Width.	226 Yards Long.	227 Yards Long.	228 Yards Long.	229 Yards Long.	230 Yards Long.
Yards.	*A. R. P.*	*A. R. P.*	*A. R. P.*	*A. R. P.*	*A. R. P.*
1	0 0 7	0 0 8	0 0 8	0 0 8	0 0 8
2	0 0 15	0 0 15	0 0 15	0 0 15	0 0 15
3	0 0 22	0 0 23	0 0 23	0 0 23	0 0 23
4	0 0 30	0 0 30	0 0 30	0 0 30	0 0 30
5	0 0 37	0 0 38	0 0 38	0 0 38	0 0 38
6	0 1 5	0 1 5	0 1 5	0 1 5	0 1 6
7	0 1 12	0 1 13	0 1 13	0 1 13	0 1 13
8	0 1 20	0 1 20	0 1 20	0 1 21	0 1 21
9	0 1 27	0 1 28	0 1 28	0 1 28	0 1 28
10	0 1 35	0 1 35	0 1 35	0 1 36	0 1 36
20	0 3 29	0 3 30	0 3 31	0 3 31	0 3 32
30	1 1 24	1 1 25	1 1 26	1 1 27	1 1 28
40	1 3 19	1 3 20	1 3 21	1 3 23	1 3 24
50	2 1 14	2 1 15	2 1 17	2 1 19	2 1 20
60	2 3 8	2 3 10	2 3 12	2 3 14	2 3 16
70	3 1 3	3 1 5	3 1 8	3 1 10	3 1 12
80	3 2 38	3 3 0	3 3 3	3 3 6	3 3 8
90	4 0 32	4 0 35	4 0 38	4 1 1	4 1 4
100	4 2 27	4 2 30	4 2 34	4 2 37	4 3 0
200	9 1 14	9 1 21	9 1 27	9 1 34	9 2 1
300	14 0 1	14 0 11	14 0 21	14 0 31	14 1 1
400	18 2 28	18 3 2	18 3 15	18 3 28	19 0 1
500	23 1 16	23 1 32	23 2 9	23 2 25	23 3 2

From 231 to 235 Yards Long.

Width.	231 Yards Long.	232 Yards Long.	233 Yards Long.	234 Yards Long.	235 Yards Long.
Yards.	*A. R. P.*	*A. R. P.*	*A. R. P.*	*A. R. P.*	*A. R. P.*
1	0 0 8	0 0 8	0 0 8	0 0 8	0 0 8
2	0 0 15	0 0 15	0 0 15	0 0 15	0 0 16
3	0 0 23	0 0 23	0 0 23	0 0 23	0 0 23
4	0 0 31	0 0 31	0 0 31	0 0 31	0 0 31
5	0 0 38	0 0 38	0 0 39	0 0 39	0 0 39
6	0 1 6	0 1 6	0 1 6	0 1 6	0 1 7
7	0 1 13	0 1 14	0 1 14	0 1 14	0 1 14
8	0 1 21	0 1 21	0 1 22	0 1 22	0 1 22
9	0 1 29	0 1 29	0 1 29	0 1 30	0 1 30
10	0 1 36	0 1 37	0 1 37	0 1 37	0 1 38
20	0 3 33	0 3 33	0 3 34	0 3 35	0 3 35
30	1 1 29	1 1 30	1 1 31	1 1 32	1 1 33
40	1 3 25	1 3 27	1 3 28	1 3 29	1 3 31
50	2 1 22	2 1 23	2 1 25	2 1 27	2 1 28
60	2 3 18	2 3 20	2 3 22	2 3 24	2 3 26
70	3 1 15	3 1 17	3 1 19	3 1 21	3 1 24
80	3 3 11	3 3 14	3 3 16	3 3 19	3 3 21
90	4 1 7	4 1 10	4 1 13	4 1 16	4 1 19
100	4 3 4	4 3 7	4 3 10	4 3 14	4 3 17
200	9 2 7	9 2 14	9 2 20	9 2 27	9 2 34
300	14 1 11	14 1 21	14 1 31	14 2 1	14 2 11
400	19 0 15	19 0 28	19 1 1	19 1 14	19 1 27
500	23 3 18	23 3 35	24 0 11	24 0 28	24 1 4

From 236 to 240 Yards Long.

Width.	236 Yards Long.	237 Yards Long.	238 Yards Long.	239 Yards Long.	240 Yards Long.
Yards.	*A. R. P.*	*A. R. P.*	*A. R. P.*	*A. R. P.*	*A. R. P.*
1	0 0 8	0 0 8	0 0 8	0 0 8	0 0 8
2	0 0 16	0 0 16	0 0 16	0 0 16	0 0 16
3	0 0 23	0 0 24	0 0 24	0 0 24	0 0 24
4	0 0 31	0 0 31	0 0 31	0 0 32	0 0 32
5	0 0 39	0 0 39	0 0 39	0 1 0	0 1 0
6	0 1 7	0 1 7	0 1 7	0 1 7	0 1 8
7	0 1 15	0 1 15	0 1 15	0 1 15	0 1 16
8	0 1 22	0 1 23	0 1 23	0 1 23	0 1 23
9	0 1 30	0 1 31	0 1 31	0 1 31	0 1 31
10	0 1 38	0 1 38	0 1 39	0 1 39	0 1 39
20	0 3 36	0 3 37	0 3 37	0 3 38	0 3 39
30	1 1 34	1 1 35	1 1 36	1 1 37	1 1 38
40	1 3 32	1 3 33	1 3 35	1 3 36	1 3 37
50	2 1 30	2 1 32	2 1 33	2 1 35	2 1 37
60	2 3 28	2 3 30	2 3 32	2 3 34	2 3 36
70	3 1 26	3 1 28	3 1 31	3 1 33	3 1 35
80	3 3 24	3 3 27	3 3 29	3 3 32	3 3 35
90	4 1 22	4 1 25	4 1 28	4 1 31	4 1 34
100	4 3 20	4 3 23	4 3 27	4 3 30	4 3 33
200	9 3 0	9 3 7	9 3 14	9 3 20	9 3 27
300	14 2 20	14 2 30	14 3 0	14 3 10	14 3 20
400	19 2 1	19 2 14	19 2 27	19 3 0	19 3 14
500	24 1 21	24 1 37	24 2 14	24 2 30	24 3 7

From 241 to 245 Yards Long.

Width.	241 Yards Long.	242 Yards Long.	243 Yards Long.	244 Yards Long.	245 Yards Long.
Yards.	*A. R. P.*	*A. R. P.*	*A. R. P.*	*A. R. P.*	*A. R. P.*
1	0 0 8	0 0 8	0 0 8	0 0 8	0 0 8
2	0 0 16	0 0 16	0 0 16	0 0 16	0 0 16
3	0 0 24	0 0 24	0 0 24	0 0 24	0 0 24
4	0 0 32	0 0 32	0 0 32	0 0 32	0 0 32
5	0 1 0	0 1 0	0 1 0	0 1 0	0 1 0
6	0 1 8	0 1 8	0 1 8	0 1 8	0 1 9
7	0 1 16	0 1 16	0 1 16	0 1 16	0 1 17
8	0 1 24	0 1 24	0 1 24	0 1 25	0 1 25
9	0 1 32	0 1 32	0 1 32	0 1 33	0 1 33
10	0 2 0	0 2 0	0 2 0	0 2 1	0 2 1
20	0 3 39	1 0 0	1 0 1	1 0 1	1 0 2
30	1 1 39	1 2 0	1 2 1	1 2 2	1 2 3
40	1 3 39	2 0 0	2 0 1	2 0 3	2 0 4
50	2 1 38	2 2 0	2 2 2	2 2 3	2 2 5
60	2 3 38	3 0 0	3 0 2	3 0 4	3 0 6
70	3 1 38	3 2 0	3 2 2	3 2 5	3 2 7
80	3 3 37	4 0 0	4 0 3	4 0 5	4 0 8
90	4 1 37	4 2 0	4 2 3	4 2 6	4 2 9
100	4 3 37	5 0 0	5 0 3	5 0 7	5 0 10
200	9 3 33	10 0 0	10 0 7	10 0 13	10 0 20
300	14 3 30	15 0 0	15 0 10	15 0 20	15 0 30
400	19 3 27	20 0 0	20 0 13	20 0 26	20 1 0
500	24 3 23	25 0 0	25 0 17	25 0 33	25 1 10

From 246 to 250 Yards Long.

Width.	246 Yards Long.	247 Yards Long.	248 Yards Long.	249 Yards Long.	250 Yards Long.
Yards.	*A. R. P.*	*A. R. P.*	*A. R. P.*	*A. R. P.*	*A. R. P.*
1	0 0 8	0 0 8	0 0 8	0 0 8	0 0 8
2	0 0 16	0 0 16	0 0 16	0 0 16	0 0 17
3	0 0 24	0 0 24	0 0 25	0 0 25	0 0 25
4	0 0 33	0 0 33	0 0 33	0 0 33	0 0 33
5	0 1 1	0 1 1	0 1 1	0 1 1	0 1 1
6	0 1 9	0 1 9	0 1 9	0 1 9	0 1 10
7	0 1 17	0 1 17	0 1 17	0 1 18	0 1 18
8	0 1 25	0 1 25	0 1 26	0 1 26	0 1 26
9	0 1 33	0 1 33	0 1 34	0 1 34	0 1 34
10	0 2 1	0 2 2	0 2 2	0 2 2	0 2 3
20	1 0 3	1 0 3	1 0 4	1 0 5	1 0 5
30	1 2 4	1 2 5	1 2 6	1 2 7	1 2 8
40	2 0 5	2 0 7	2 0 8	2 0 9	2 0 11
50	2 2 7	2 2 8	2 2 10	2 2 12	2 2 13
60	3 0 8	3 0 10	3 0 12	3 0 14	3 0 16
70	3 2 9	3 2 12	3 2 14	3 2 16	3 2 19
80	4 0 11	4 0 13	4 0 16	4 0 19	4 0 21
90	4 2 12	4 2 15	4 2 18	4 2 21	4 2 24
100	5 0 13	5 0 17	5 0 20	5 0 23	5 0 26
200	10 0 26	10 0 33	10 1 0	10 1 6	10 1 13
300	15 1 0	15 1 10	15 1 20	15 1 29	15 1 39
400	20 1 13	20 1 26	20 1 39	20 2 13	20 2 26
500	25 1 26	25 2 3	25 2 19	25 2 36	25 3 12

From 251 to 255 Yards Long.

Width.	251 Yards Long.	252 Yards Long.	253 Yards Long.	254 Yards Long.	255 Yards Long
Yards.	*A. R. P.*	*A. R. P.*	*A. R. P.*	*A. R. P.*	*A. R. P.*
1	0 0 8	0 0 8	0 0 8	0 0 8	0 0 8
2	0 0 17	0 0 17	0 0 17	0 0 17	0 0 17
3	0 0 25	0 0 25	0 0 25	0 0 25	0 0 25
4	0 0 33	0 0 33	0 0 33	0 0 34	0 0 34
5	0 1 1	0 1 2	0 1 2	0 1 2	0 1 2
6	0 1 10	0 1 10	0 1 10	0 1 10	0 1 11
7	0 1 18	0 1 18	0 1 19	0 1 19	0 1 19
8	0 1 26	0 1 27	0 1 27	0 1 27	0 1 27
9	0 1 35	0 1 35	0 1 35	0 1 36	0 1 36
10	0 2 3	0 2 3	0 2 4	0 2 4	0 2 4
20	1 0 6	1 0 7	1 0 7	1 0 8	1 0 9
30	1 2 9	1 2 10	1 2 11	1 2 12	1 2 13
40	2 0 12	2 0 13	2 0 15	2 0 16	2 0 17
50	2 2 15	2 2 17	2 2 18	2 2 20	2 2 21
60	3 0 18	3 0 20	3 0 22	3 0 24	3 0 26
70	3 2 21	3 2 23	3 2 25	3 2 28	3 2 30
80	4 0 24	4 0 26	4 0 29	4 0 32	4 0 34
90	4 2 27	4 2 30	4 2 33	4 2 36	4 2 39
100	5 0 30	5 0 33	5 0 36	5 1 0	5 1 3
200	10 1 20	10 1 26	10 1 33	10 1 39	10 2 6
300	15 2 9	15 2 19	15 2 29	15 2 39	15 3 9
400	20 2 39	20 3 12	20 3 25	20 3 39	21 0 12
500	25 3 29	26 0 5	26 0 22	26 0 38	26 1 15

From 256 to 260 Yards Long.

Width.	256 Yards Long.	257 Yards Long.	258 Yards Long.	259 Yards Long.	260 Yards Long.
Yards.	*A. R. P.*	*A. R. P.*	*A. R. P.*	*A. R. P.*	*A. R. P.*
1	0 0 8	0 0 8	0 0 9	0 0 9	0 0 9
2	0 0 17	0 0 17	0 0 17	0 0 17	0 0 17
3	0 0 25	0 0 25	0 0 26	0 0 26	0 0 26
4	0 0 34	0 0 34	0 0 34	0 0 34	0 0 34
5	0 1 2	0 1 2	0 1 3	0 1 3	0 1 3
6	0 1 11	0 1 11	0 1 11	0 1 11	0 1 12
7	0 1 19	0 1 19	0 1 20	0 1 20	0 1 20
8	0 1 28	0 1 28	0 1 28	0 1 28	0 1 29
9	0 1 36	0 1 36	0 1 37	0 1 37	0 1 37
10	0 2 5	0 2 5	0 2 5	0 2 6	0 2 6
20	1 0 9	1 0 10	1 0 11	1 0 11	1 0 12
30	1 2 14	1 2 15	1 2 16	1 2 17	1 2 18
40	2 0 19	2 0 20	2 0 21	2 0 22	2 0 24
50	2 2 23	2 2 25	2 2 26	2 2 28	2 2 30
60	3 0 28	3 0 30	3 0 32	3 0 34	3 0 36
70	3 2 32	3 2 35	3 2 37	3 2 39	3 3 2
80	4 0 37	4 1 0	4 1 2	4 1 5	4 1 8
90	4 3 2	4 3 5	4 3 8	4 3 11	4 3 14
100	5 1 6	5 1 10	5 1 13	5 1 16	5 1 20
200	10 2 13	10 2 19	10 2 26	10 2 32	10 2 39
300	15 3 19	15 3 29	15 3 39	16 0 9	16 0 19
400	21 0 25	21 0 38	21 1 12	21 1 25	21 1 38
500	26 1 31	26 2 8	26 2 24	26 3 1	26 3 18

From 261 to 265 Yards Long.

Width.	261 Yards Long.	262 Yards Long.	263 Yards Long.	264 Yards Long.	265 Yards Long.
Yards.	*A. R. P.*	*A. R. P.*	*A. R. P.*	*A. R. P.*	*A. R. P.*
1	0 0 9	0 0 9	0 0 9	0 0 9	0 0 9
2	0 0 17	0 0 17	0 0 17	0 0 17	0 0 18
3	0 0 26	0 0 26	0 0 26	0 0 26	0 0 26
4	0 0 35	0 0 35	0 0 35	0 0 35	0 0 35
5	0 1 3	0 1 3	0 1 3	0 1 4	0 1 4
6	0 1 12	0 1 12	0 1 12	0 1 12	0 1 13
7	0 1 20	0 1 21	0 1 21	0 1 21	0 1 21
8	0 1 29	0 1 29	0 1 30	0 1 30	0 1 30
9	0 1 38	0 1 38	0 1 38	0 1 39	0 1 39
10	0 2 6	0 2 7	0 2 7	0 2 7	0 2 8
20	1 0 13	1 0 13	1 0 14	1 0 15	1 0 15
30	1 2 19	1 2 20	1 2 21	1 2 22	1 2 23
40	2 0 25	2 0 26	2 0 28	2 0 29	2 0 30
50	2 2 31	2 2 33	2 2 35	2 2 36	2 2 38
60	3 0 38	3 1 0	3 1 2	3 1 4	3 1 6
70	3 3 4	3 3 6	3 3 9	3 3 11	3 3 13
80	4 1 10	4 1 13	4 1 16	4 1 18	4 1 21
90	4 3 17	4 3 20	4 3 22	4 3 25	4 3 28
100	5 1 23	5 1 26	5 1 29	5 1 33	5 1 36
200	10 3 6	10 3 12	10 3 19	10 3 25	10 3 32
300	16 0 28	16 0 38	16 1 8	16 1 18	16 1 28
400	21 2 11	21 2 24	21 2 38	21 3 11	21 3 24
500	26 3 34	27 0 11	27 0 27	27 1 4	27 1 20

From 266 to 270 Yards Long.

Width.	266 Yards Long.	267 Yards Long.	268 Yards Long.	269 Yards Long.	270 Yards Long.
Yards.	*A. R. P.*	*A. R. P.*	*A. R. P.*	*A. R. P.*	*A. R. P.*
1	0 0 9	0 0 9	0 0 9	0 0 9	0 0 9
2	0 0 18	0 0 18	0 0 18	0 0 18	0 0 18
3	0 0 26	0 0 26	0 0 27	0 0 27	0 0 27
4	0 0 35	0 0 35	0 0 35	0 0 36	0 0 36
5	0 1 4	0 1 4	0 1 4	0 1 4	0 1 5
6	0 1 13	0 1 13	0 1 13	0 1 13	0 1 14
7	0 1 22	0 1 22	0 1 22	0 1 22	0 1 22
8	0 1 30	0 1 31	0 1 31	0 1 31	0 1 31
9	0 1 39	0 1 39	0 2 0	0 2 0	0 2 0
10	0 2 8	0 2 8	0 2 9	0 2 9	0 2 9
20	1 0 16	1 0 17	1 0 17	1 0 18	1 0 19
30	1 2 24	1 2 25	1 2 26	1 2 27	1 2 28
40	2 0 32	2 0 33	2 0 34	2 0 36	2 0 37
50	2 3 0	2 3 1	2 3 3	2 3 5	2 3 6
60	3 1 8	3 1 10	3 1 12	3 1 14	3 1 16
70	3 3 16	3 3 18	3 3 20	3 3 22	3 3 25
80	4 1 23	4 1 26	4 1 29	4 1 31	4 1 34
90	4 3 31	4 3 34	4 3 37	5 0 0	5 0 3
100	5 1 39	5 2 3	5 2 6	5 2 9	5 2 13
200	10 3 39	11 0 5	11 0 12	11 0 19	11 0 25
300	16 1 38	16 2 8	16 2 18	16 2 28	16 2 38
400	21 3 37	22 0 11	22 0 24	22 0 37	22 1 10
500	27 1 37	27 2 13	27 2 30	27 3 6	27 3 23

From 271 to 275 Yards Long.

Width.	271 Yards Long.	272 Yards Long.	273 Yards Long.	274 Yards Long.	275 Yards Long.
Yards.	*A. R. P.*	*A. R. P.*	*A. R. P.*	*A. R. P.*	*A. R. P.*
1	0 0 9	0 0 9	0 0 9	0 0 9	0 0 9
2	0 0 18	0 0 18	0 0 18	0 0 18	0 0 18
3	0 0 27	0 0 27	0 0 27	0 0 27	0 0 27
4	0 0 36	0 0 36	0 0 36	0 0 36	0 0 36
5	0 1 5	0 1 5	0 1 5	0 1 5	0 1 5
6	0 1 14	0 1 14	0 1 14	0 1 14	0 1 15
7	0 1 23	0 1 23	0 1 23	0 1 23	0 1 24
8	0 1 32	0 1 32	0 1 32	0 1 32	0 1 33
9	0 2 1	0 2 1	0 2 1	0 2 2	0 2 2
10	0 2 10	0 2 10	0 2 10	0 2 11	0 2 11
20	1 0 19	1 0 20	1 0 20	1 0 21	1 0 22
30	1 2 29	1 2 30	1 2 31	1 2 32	1 2 33
40	2 0 38	2 1 0	2 1 1	2 1 2	2 1 4
50	2 3 8	2 3 10	2 3 11	2 3 13	2 3 15
60	3 1 18	3 1 20	3 1 21	3 1 23	3 1 25
70	3 3 27	3 3 29	3 3 32	3 3 34	3 3 36
80	4 1 37	4 1 39	4 2 2	4 2 5	4 2 7
90	5 0 6	5 0 9	5 0 12	5 0 15	5 0 18
100	5 2 16	5 2 19	5 2 22	5 2 26	5 2 29
200	11 0 32	11 0 38	11 1 5	11 1 12	11 1 18
300	16 3 8	16 3 18	16 3 27	16 3 37	17 0 7
400	22 1 23	22 1 37	22 2 10	22 2 23	22 2 36
500	27 3 39	28 0 16	28 0 32	28 1 9	28 1 25

From 276 to 280 Yards Long.

Width.	276 Yards Long.	277 Yards Long.	278 Yards Long.	279 Yards Long.	280 Yards Long.
Yards.	*A. R. P.*	*A. R. P.*	*A. R. P.*	*A. R. P.*	*A. R. P.*
1	0 0 9	0 0 9	0 0 9	0 0 9	0 0 9
2	0 0 18	0 0 18	0 0 18	0 0 18	0 0 19
3	0 0 27	0 0 27	0 0 28	0 0 28	0 0 28
4	0 0 36	0 0 37	0 0 37	0 0 37	0 0 37
5	0 1 6	0 1 6	0 1 6	0 1 6	0 1 6
6	0 1 15	0 1 15	0 1 15	0 1 15	0 1 16
7	0 1 24	0 1 24	0 1 24	0 1 25	0 1 25
8	0 1 33	0 1 33	0 1 34	0 1 34	0 1 34
9	0 2 2	0 2 2	0 2 3	0 2 3	0 2 3
10	0 2 11	0 2 12	0 2 12	0 2 12	0 2 13
20	1 0 22	1 0 23	1 0 24	1 0 24	1 0 25
30	1 2 34	1 2 35	1 2 36	1 2 37	1 2 38
40	2 1 5	2 1 6	2 1 8	2 1 9	2 1 10
50	2 3 16	2 3 18	2 3 20	2 3 21	2 3 23
60	3 1 27	3 1 29	3 1 31	3 1 33	3 1 35
70	3 3 39	4 0 1	4 0 3	4 0 6	4 0 8
80	4 2 10	4 2 13	4 2 15	4 2 18	4 2 20
90	5 0 21	5 0 24	5 0 27	5 0 30	5 0 33
100	5 2 32	5 2 36	5 2 39	5 3 2	5 3 6
200	11 1 25	11 1 31	11 1 38	11 2 5	11 2 11
300	17 0 17	17 0 27	17 0 37	17 1 7	17 1 17
400	22 3 10	22 3 23	22 3 36	23 0 9	23 0 22
500	28 2 2	28 2 19	28 2 35	28 3 12	28 3 28

From 281 to 285 Yards Long.

Width.	281 Yards Long.	282 Yards Long.	283 Yards Long.	284 Yards Long.	285 Yards Long.
Yards.	*A. R. P.*	*A. R. P.*	*A. R. P.*	*A. R. P.*	*A. R. P.*
1	0 0 9	0 0 9	0 0 9	0 0 9	0 0 9
2	0 0 19	0 0 19	0 0 19	0 0 19	0 0 19
3	0 0 28	0 0 28	0 0 28	0 0 28	0 0 28
4	0 0 37	0 0 37	0 0 37	0 0 38	0 0 38
5	0 1 6	0 1 7	0 1 7	0 1 7	0 1 7
6	0 1 16	0 1 16	0 1 16	0 1 16	0 1 17
7	0 1 25	0 1 25	0 1 25	0 1 26	0 1 26
8	0 1 34	0 1 35	0 1 35	0 1 35	0 1 35
9	0 2 4	0 2 4	0 2 4	0 2 4	0 2 5
10	0 2 13	0 2 13	0 2 14	0 2 14	0 2 14
20	1 0 26	1 0 26	1 0 27	1 0 28	1 0 28
30	1 2 39	1 3 0	1 3 1	1 3 2	1 3 3
40	2 1 12	2 1 13	2 1 14	2 1 16	2 1 17
50	2 3 24	2 3 26	2 3 28	2 3 29	2 3 31
60	3 1 37	3 1 39	3 2 1	3 2 3	3 2 5
70	4 0 10	4 0 13	4 0 15	4 0 17	4 0 20
80	4 2 23	4 2 26	4 2 28	4 2 31	4 2 34
90	5 0 36	5 0 39	5 1 2	5 1 5	5 1 8
100	5 3 9	5 3 12	5 3 16	5 3 19	5 3 22
200	11 2 18	11 2 24	11 2 31	11 2 38	11 3 4
300	17 1 27	17 1 37	17 2 7	17 2 17	17 2 26
400	23 0 36	23 1 9	23 1 22	23 1 35	23 2 9
500	29 0 5	29 0 21	29 0 38	29 1 14	29 1 31

From 286 to 290 Yards Long.

Width.	286 Yards Long.	287 Yards Long.	288 Yards Long.	289 Yards Long.	290 Yards Long.
Yards.	*A. R. P.*	*A. R. P.*	*A. R. P.*	*A. R. P.*	*A. R. P.*
1	0 0 9	0 0 9	0 0 10	0 0 10	0 0 10
2	0 0 19	0 0 19	0 0 19	0 0 19	0 0 19
3	0 0 28	0 0 28	0 0 29	0 0 29	0 0 29
4	0 0 38	0 0 38	0 0 38	0 0 38	0 0 38
5	0 1 7	0 1 7	0 1 8	0 1 8	0 1 8
6	0 1 17	0 1 17	0 1 17	0 1 17	0 1 18
7	0 1 26	0 1 26	0 1 27	0 1 27	0 1 27
8	0 1 36	0 1 36	0 1 36	0 1 36	0 1 37
9	0 2 5	0 2 5	0 2 6	0 2 6	0 2 6
10	0 2 15	0 2 15	0 2 15	0 2 16	0 2 16
20	1 0 29	1 0 30	1 0 30	1 0 31	1 0 32
30	1 3 4	1 3 5	1 3 6	1 3 7	1 3 8
40	2 1 18	2 1 20	2 1 21	2 1 22	2 1 23
50	2 3 33	2 3 34	2 3 36	2 3 38	2 3 39
60	3 2 7	3 2 9	3 2 11	3 2 13	3 2 15
70	4 0 22	4 0 24	4 0 26	4 0 29	4 0 31
80	4 2 36	4 2 39	4 3 2	4 3 4	4 3 7
90	5 1 11	5 1 14	5 1 17	5 1 20	5 1 23
100	5 3 25	5 3 29	5 3 32	5 3 35	5 3 39
200	11 3 11	11 3 18	11 3 24	11 3 31	11 3 37
300	17 2 36	17 3 6	17 3 16	17 3 26	17 3 36
400	23 2 22	23 2 35	23 3 8	23 3 21	23 3 35
500	29 2 7	29 2 24	29 3 0	29 3 17	29 3 33

From 291 to 295 Yards Long.

Width.	291 Yards Long.	292 Yards Long.	293 Yards Long.	294 Yards Long.	295 Yards Long.
Yards.	*A. R. P.*	*A. R. P.*	*A. R. P.*	*A. R. P.*	*A. R. P.*
1	0 0 10	0 0 10	0 0 10	0 0 10	0 0 10
2	0 0 19	0 0 19	0 0 19	0 0 19	0 0 20
3	0 0 29	0 0 29	0 0 29	0 0 29	0 0 29
4	0 0 38	0 0 39	0 0 39	0 0 39	0 0 39
5	0 1 8	0 1 8	0 1 8	0 1 9	0 1 9
6	0 1 18	0 1 18	0 1 18	0 1 18	0 1 19
7	0 1 27	0 1 28	0 1 28	0 1 28	0 1 28
8	0 1 37	0 1 37	0 1 37	0 1 38	0 1 38
9	0 2 7	0 2 7	0 2 7	0 2 7	0 2 8
10	0 2 16	0 2 17	0 2 17	0 2 17	0 2 18
20	1 0 32	1 0 33	1 0 34	1 0 34	1 0 35
30	1 3 9	1 3 10	1 3 11	1 3 12	1 3 13
40	2 1 25	2 1 26	2 1 27	2 1 29	2 1 30
50	3 0 1	3 0 3	3 0 4	3 0 6	3 0 8
60	3 2 17	3 2 19	3 2 21	3 2 23	3 2 25
70	4 0 33	4 0 36	4 0 38	4 1 0	4 1 3
80	4 3 10	4 3 12	4 3 15	4 3 18	4 3 20
90	5 1 26	5 1 29	5 1 32	5 1 35	5 1 38
100	6 0 2	6 0 5	6 0 9	6 0 12	6 0 15
200	12 0 4	12 0 11	12 0 17	12 0 24	12 0 30
300	18 0 6	18 0 16	18 0 26	18 0 36	18 1 6
400	24 0 8	24 0 21	24 0 34	24 1 8	24 1 21
500	30 0 10	30 0 26	30 1 3	30 1 20	30 1 36

From 296 to 300 Yards Long.

Width.	296 Yards Long.	297 Yards Long.	298 Yards Long.	299 Yards Long.	300 Yards Long.
Yards.	*A. R. P.*	*A. R. P.*	*A. R. P.*	*A. R. P.*	*A. R. P.*
1	0 0 10	0 0 10	0 0 10	0 0 10	0 0 10
2	0 0 20	0 0 20	0 0 20	0 0 20	0 0 20
3	0 0 29	0 0 29	0 0 30	0 0 30	0 0 30
4	0 0 39	0 0 39	0 0 39	0 1 0	0 1 0
5	0 1 9	0 1 9	0 1 9	0 1 9	0 1 10
6	0 1 19	0 1 19	0 1 19	0 1 19	0 1 20
7	0 1 28	0 1 29	0 1 29	0 1 29	0 1 29
8	0 1 38	0 1 39	0 1 39	0 1 39	0 1 39
9	0 2 8	0 2 8	0 2 9	0 2 9	0 2 9
10	0 2 18	0 2 18	0 2 19	0 2 19	0 2 19
20	1 0 36	1 0 36	1 0 37	1 0 38	1 0 38
30	1 3 14	1 3 15	1 3 16	1 3 17	1 3 18
40	2 1 31	2 1 33	2 1 34	2 1 35	2 1 37
50	3 0 9	3 0 11	3 0 13	3 0 14	3 0 16
60	3 2 27	3 2 29	3 2 31	3 2 33	3 2 35
70	4 1 5	4 1 7	4 1 10	4 1 12	4 1 14
80	4 3 23	4 3 25	4 3 28	4 3 31	4 3 33
90	5 2 1	5 2 4	5 2 7	5 2 10	5 2 13
100	6 0 19	6 0 22	6 0 25	6 0 28	6 0 32
200	12 0 37	12 1 4	12 1 10	12 1 17	12 1 23
300	18 1 16	18 1 25	18 1 35	18 2 5	18 2 15
400	24 1 34	24 2 7	24 2 20	24 2 34	24 3 7
500	30 2 13	30 2 29	30 3 6	30 3 22	30 3 39

From 301 to 305 Yards Long.

Width.	301 Yards Long.	302 Yards Long.	303 Yards Long.	304 Yards Long.	305 Yards Long.
Yards.	*A. R. P.*	*A. R. P.*	*A. R. P.*	*A. R. P.*	*A. R. P.*
1	0 0 10	0 0 10	0 0 10	0 0 10	0 0 10
2	0 0 20	0 0 20	0 0 20	0 0 20	0 0 20
3	0 0 30	0 0 30	0 0 30	0 0 30	0 0 30
4	0 1 0	0 1 0	0 1 0	0 1 0	0 1 0
5	0 1 10	0 1 10	0 1 10	0 1 10	0 1 10
6	0 1 20	0 1 20	0 1 20	0 1 20	0 1 20
7	0 1 30	0 1 30	0 1 30	0 1 30	0 1 31
8	0 2 0	0 2 0	0 2 0	0 2 0	0 2 1
9	0 2 10	0 2 10	0 2 10	0 2 10	0 2 11
10	0 2 20	0 2 20	0 2 20	0 2 20	0 2 21
20	1 0 39	1 1 0	1 1 0	1 1 1	1 1 2
30	1 3 19	1 3 20	1 3 20	1 3 21	1 3 22
40	2 1 38	2 1 39	2 2 1	2 2 2	2 2 3
50	3 0 18	3 0 19	3 0 21	3 0 22	3 0 24
60	3 2 37	3 2 39	3 3 1	3 3 3	3 3 5
70	4 1 17	4 1 19	4 1 21	4 1 23	4 1 26
80	4 3 36	4 3 39	5 0 1	5 0 4	5 0 7
90	5 2 16	5 2 19	5 2 21	5 2 24	5 2 27
100	6 0 35	6 0 38	6 1 2	6 1 5	6 1 8
200	12 1 30	12 1 37	12 2 3	12 2 10	12 2 17
300	18 2 25	18 2 35	18 3 5	18 3 15	18 3 25
400	24 3 20	24 3 33	25 0 7	25 0 20	25 0 33
500	31 0 15	31 0 32	31 1 8	31 1 25	31 2 1

From 306 to 310 Yards Long.

Width.	306 Yards Long.	307 Yards Long.	308 Yards Long.	309 Yards Long.	310 Yards Long.
Yards.	*A. R. P.*	*A. R. P.*	*A. R. P.*	*A. R. P.*	*A. R. P.*
1	0 0 10	0 0 10	0 0 10	0 0 10	0 0 10
2	0 0 20	0 0 20	0 0 20	0 0 20	0 0 20
3	0 0 30	0 0 30	0 0 31	0 0 31	0 0 31
4	0 1 0	0 1 1	0 1 1	0 1 1	0 1 1
5	0 1 11	0 1 11	0 1 11	0 1 11	0 1 11
6	0 1 21	0 1 21	0 1 21	0 1 21	0 1 21
7	0 1 31	0 1 31	0 1 31	0 1 32	0 1 32
8	0 2 1	0 2 1	0 2 1	0 2 2	0 2 2
9	0 2 11	0 2 11	0 2 12	0 2 12	0 2 12
10	0 2 21	0 2 21	0 2 22	0 2 22	0 2 22
20	1 1 2	1 1 3	1 1 4	1 1 4	1 1 5
30	1 3 23	1 3 24	1 3 25	1 3 26	1 3 27
40	2 2 5	2 2 6	2 2 7	2 2 9	2 2 10
50	3 0 26	3 0 27	3 0 29	3 0 31	3 0 32
60	3 3 7	3 3 9	3 3 11	3 3 13	3 3 15
70	4 1 28	4 1 30	4 1 33	4 1 35	4 1 37
80	5 0 9	5 0 12	5 0 15	5 0 17	5 0 20
90	5 2 30	5 2 33	5 2 36	5 2 39	5 3 2
100	6 1 12	6 1 15	6 1 18	6 1 21	6 1 25
200	12 2 23	12 2 30	12 2 36	12 3 3	12 3 10
300	18 3 35	19 0 5	19 0 15	19 0 24	19 0 34
400	25 1 6	25 1 20	25 1 33	25 2 6	25 2 19
500	31 2 18	31 2 34	31 3 11	31 3 27	32 0 4

From 311 to 315 Yards Long.

Width.	311 Yards Long.	312 Yards Long.	313 Yards Long.	314 Yards Long.	315 Yards Long.
Yards.	*A. R. P.*	*A. R. P.*	*A. R. P.*	*A. R. P.*	*A. R. P.*
1	0 0 10	0 0 10	0 0 10	0 0 10	0 0 10
2	0 0 21	0 0 21	0 0 21	0 0 21	0 0 21
3	0 0 31	0 0 31	0 0 31	0 0 31	0 0 31
4	0 1 1	0 1 1	0 1 1	0 1 2	0 1 2
5	0 1 11	0 1 12	0 1 12	0 1 12	0 1 12
6	0 1 22	0 1 22	0 1 22	0 1 22	0 1 22
7	0 1 32	0 1 32	0 1 32	0 1 33	0 1 33
8	0 2 2	0 2 3	0 2 3	0 2 3	0 2 3
9	0 2 13	0 2 13	0 2 13	0 2 13	0 2 14
10	0 2 23	0 2 23	0 2 23	0 2 24	0 2 24
20	1 1 6	1 1 6	1 1 7	1 1 8	1 1 8
30	1 3 28	1 3 29	1 3 30	1 3 31	1 3 32
40	2 2 11	2 2 13	2 2 14	2 2 15	2 2 17
50	3 0 34	3 0 36	3 0 37	3 0 39	3 1 1
60	3 3 17	3 3 19	3 3 21	3 3 23	3 3 25
70	4 2 0	4 2 2	4 2 4	4 2 7	4 2 9
80	5 0 22	5 0 25	5 0 28	5 0 30	5 0 33
90	5 3 5	5 3 8	5 3 11	5 3 14	5 3 17
100	6 0 28	6 1 31	6 1 35	6 1 38	6 2 1
200	12 1 16	12 3 23	12 3 29	12 3 36	13 0 3
300	19 1 4	19 1 14	19 1 24	19 1 34	19 2 4
400	25 2 32	25 3 6	25 3 19	25 3 32	26 0 5
500	32 0 20	32 0 37	32 1 14	32 1 30	32 2 7

From 316 to 320 Yards Long.

Width.	316 Yards Long.	317 Yards Long.	318 Yards Long.	319 Yards Long.	320 Yards Long.
Yards.	*A. R. P.*	*A. R. P.*	*A. R. P.*	*A. R. P.*	*A. R. P.*
1	0 0 10	0 0 10	0 0 11	0 0 11	0 0 11
2	0 0 21	0 0 21	0 0 21	0 0 21	0 0 21
3	0 0 31	0 0 31	0 0 32	0 0 32	0 0 32
4	0 1 2	0 1 2	0 1 2	0 1 2	0 1 2
5	0 1 12	0 1 12	0 1 13	0 1 13	0 1 13
6	0 1 23	0 1 23	0 1 23	0 1 23	0 1 23
7	0 1 33	0 1 33	0 1 34	0 1 34	0 1 34
8	0 2 4	0 2 4	0 2 4	0 2 4	0 2 5
9	0 2 14	0 2 14	0 2 15	0 2 15	0 2 15
10	0 2 24	0 2 25	0 2 25	0 2 25	0 2 26
20	1 1 9	1 1 10	1 1 10	1 1 11	1 1 12
30	1 3 33	1 3 34	1 3 35	1 3 36	1 3 37
40	2 2 18	2 2 19	2 2 20	2 2 22	2 2 23
50	3 1 2	3 1 4	3 1 6	3 1 7	3 1 9
60	3 3 27	3 3 29	3 3 31	3 3 33	3 3 35
70	4 2 11	4 2 14	4 2 16	4 2 18	4 2 20
80	5 0 36	5 0 38	5 1 1	5 1 4	5 1 6
90	5 3 20	5 3 23	5 3 26	5 3 29	5 3 32
100	6 2 5	6 2 8	6 2 11	6 2 15	6 2 18
200	13 0 9	13 0 16	13 0 22	13 0 29	13 0 36
300	19 2 14	19 2 24	19 2 34	19 3 4	19 3 14
400	26 0 19	26 0 32	26 1 5	26 1 18	26 1 31
500	32 2 23	32 3 0	32 3 16	32 3 33	33 0 9

From 321 to 325 Yards Long.

Width.	321 Yards Long.	322 Yards Long.	323 Yards Long.	324 Yards Long.	325 Yards Long.
Yards.	*A. R. P.*	*A. R. P.*	*A. R. P.*	*A. R. P.*	*A. R. P.*
1	0 0 11	0 0 11	0 0 11	0 0 11	0 0 11
2	0 0 21	0 0 21	0 0 21	0 0 21	0 0 21
3	0 0 32	0 0 32	0 0 32	0 0 32	0 0 32
4	0 1 2	0 1 3	0 1 3	0 1 3	0 1 3
5	0 1 13	0 1 13	0 1 13	0 1 14	0 1 14
6	0 1 24	0 1 24	0 1 24	0 1 24	0 1 24
7	0 1 34	0 1 35	0 1 35	0 1 35	0 1 35
8	0 2 5	0 2 5	0 2 5	0 2 6	0 2 6
9	0 2 16	0 2 16	0 2 16	0 2 16	0 2 17
10	0 2 26	0 2 26	0 2 27	0 2 27	0 2 27
20	1 1 12	1 1 13	1 1 14	1 1 14	1 1 15
30	1 3 38	1 3 39	2 0 0	2 0 1	2 0 2
40	2 2 24	2 2 26	2 2 27	2 2 28	2 2 30
50	3 1 11	3 1 12	3 1 14	3 1 16	3 1 17
60	3 3 37	3 3 39	4 0 1	4 0 3	4 0 5
70	4 2 23	4 2 25	4 2 27	4 2 30	4 2 32
80	5 1 9	5 1 12	5 1 14	5 1 17	5 1 20
90	5 3 35	5 3 38	6 0 1	6 0 4	6 0 7
100	6 2 21	6 2 24	6 2 28	6 2 31	6 2 34
200	13 1 2	13 1 9	13 1 16	13 1 22	13 1 29
300	19 3 23	19 3 33	20 0 3	20 0 13	20 0 23
400	26 2 5	26 2 18	26 2 31	26 3 4	26 3 18
500	33 0 26	33 1 2	33 1 19	33 1 35	33 2 12

From 326 to 330 Yards Long.

Width.	326 Yards Long.	327 Yards Long.	328 Yards Long.	329 Yards Long.	330 Yards Long.
Yards.	*A. R. P.*	*A. R. P.*	*A. R. P.*	*A. R. P.*	*A. R. P.*
1	0 0 11	0 0 11	0 0 11	0 0 11	0 0 11
2	0 0 22	0 0 22	0 0 22	0 0 22	0 0 22
3	0 0 32	0 0 32	0 0 33	0 0 33	0 0 33
4	0 1 3	0 1 3	0 1 3	0 1 4	0 1 4
5	0 1 14	0 1 14	0 1 14	0 1 14	0 1 15
6	0 1 25	0 1 25	0 1 25	0 1 25	0 1 25
7	0 1 35	0 1 36	0 1 36	0 1 36	0 1 36
8	0 2 6	0 2 6	0 2 7	0 2 7	0 2 7
9	0 2 17	0 2 17	0 2 18	0 2 18	0 2 18
10	0 2 28	0 2 28	0 2 28	0 2 29	0 2 29
20	1 1 16	1 1 16	1 1 17	1 1 18	1 1 18
30	2 0 3	2 0 4	2 0 5	2 0 6	2 0 7
40	2 2 31	2 2 32	2 2 34	2 2 35	2 2 36
50	3 1 19	3 1 20	3 1 22	3 1 24	3 1 25
60	4 0 7	4 0 9	4 0 11	4 0 13	4 0 15
70	4 2 34	4 2 37	4 2 39	4 3 1	4 3 4
80	5 1 22	5 1 25	5 1 27	5 1 30	5 1 33
90	6 0 10	6 0 13	6 0 16	6 0 19	6 0 22
100	6 2 38	6 3 1	6 3 4	6 3 8	6 3 11
200	13 1 35	13 2 2	13 2 9	13 2 15	13 2 22
300	20 0 33	20 1 3	20 1 13	20 1 23	20 1 33
400	26 3 31	27 0 4	27 0 17	27 0 30	27 1 4
500	33 2 28	33 3 5	33 3 21	33 3 38	34 0 15

From 331 to 335 Yards Long.

Width.	331 Yards Long.	332 Yards Long.	333 Yards Long.	334 Yards Long.	335 Yards Long.
Yards.	*A. R. P.*	*A. R. P.*	*A. R. P.*	*A. R. P.*	*A. R. P.*
1	0 0 11	0 0 11	0 0 11	0 0 11	0 0 11
2	0 0 22	0 0 22	0 0 22	0 0 22	0 0 22
3	0 0 33	0 0 32	0 0 33	0 0 33	0 0 33
4	0 1 4	0 1 4	0 1 4	0 1 4	0 1 4
5	0 1 15	0 1 15	0 1 15	0 1 15	0 1 15
6	0 1 26	0 1 26	0 1 26	0 1 26	0 1 26
7	0 1 37	0 1 37	0 1 37	0 1 37	0 1 38
8	0 2 8	0 2 8	0 2 8	0 2 8	0 2 9
9	0 2 18	0 2 19	0 2 19	0 2 19	0 2 20
10	0 2 29	0 2 30	0 2 30	0 2 30	0 2 31
20	1 1 19	1 1 20	1 1 20	1 1 21	1 1 21
30	2 0 8	2 0 9	2 0 10	2 0 11	2 0 12
40	2 2 38	2 2 39	2 3 0	2 3 2	2 3 3
50	3 1 27	3 1 29	3 1 30	3 1 32	3 1 34
60	4 0 17	4 0 19	4 0 20	4 0 22	4 0 24
70	4 2 6	4 3 8	4 3 11	4 3 13	4 3 15
80	5 1 35	5 1 38	5 2 1	5 2 3	5 2 6
90	6 0 25	6 0 28	6 0 31	6 0 34	6 0 37
100	6 3 14	6 3 18	6 3 21	6 3 24	6 3 27
200	13 2 28	13 2 35	13 3 2	13 3 8	13 3 15
300	20 2 3	20 2 13	20 2 22	20 2 32	20 3 2
400	27 1 17	27 1 30	27 2 3	27 2 17	27 2 30
500	34 0 31	34 1 8	34 1 24	34 2 1	34 2 17

From 341 to 345 Yards Long.

Width.	341 Yards Long.	342 Yards Long.	343 Yards Long.	344 Yards Long.	345 Yards Long.
Yards.	*A. R. P.*	*A. R. P.*	*A. R. P.*	*A. R. P.*	*A. R. P.*
1	0 0 11	0 0 11	0 0 11	0 0 11	0 0 11
2	0 0 23	0 0 23	0 0 23	0 0 23	0 0 23
3	0 0 34	0 0 34	0 0 34	0 0 34	0 0 34
4	0 1 5	0 1 5	0 1 5	0 1 5	0 1 6
5	0 1 16	0 1 17	0 1 17	0 1 17	0 1 17
6	0 1 28	0 1 28	0 1 28	0 1 28	0 1 28
7	0 1 39	0 1 39	0 1 39	0 2 0	0 2 0
8	0 2 10	0 2 10	0 2 11	0 2 11	0 2 11
9	0 2 21	0 2 22	0 2 22	0 2 22	0 2 23
10	0 2 33	0 2 33	0 2 33	0 2 34	0 2 34
20	1 1 25	1 1 26	1 1 27	1 1 27	1 1 28
30	2 0 18	2 0 19	2 0 20	2 0 21	2 0 22
40	2 3 11	2 3 12	2 3 14	2 3 15	2 3 16
50	3 2 4	3 2 5	3 2 7	3 2 9	3 2 10
60	4 0 36	4 0 38	4 1 0	4 1 2	4 1 4
70	4 3 29	4 3 31	4 3 34	4 3 36	4 3 38
80	5 2 22	5 2 24	5 2 27	5 2 30	5 2 32
90	6 1 15	6 1 18	6 1 20	6 1 23	6 1 26
100	7 0 7	7 0 11	7 0 14	7 0 17	7 0 20
200	14 0 15	14 0 21	14 0 28	14 0 34	14 1 1
300	21 0 22	21 0 32	21 1 2	21 1 12	21 1 21
400	28 0 29	28 1 2	28 1 16	28 1 29	28 2 2
500	35 0 36	35 1 13	35 1 29	35 2 6	35 2 22

From 346 to 350 Yards Long.

Width.	346 Yards Long.	347 Yards Long.	348 Yards Long.	349 Yards Long.	350 Yards Long.
Yards.	*A. R. P.*	*A. R. P.*	*A. R. P.*	*A. R. P.*	*A. R. P.*
1	0 0 11	0 0 11	0 0 12	0 0 12	0 0 12
2	0 0 23	0 0 23	0 0 23	0 0 23	0 0 23
3	0 0 34	0 0 34	0 0 35	0 0 35	0 0 35
4	0 1 6	0 1 6	0 1 6	0 1 6	0 1 6
5	0 1 17	0 1 17	0 1 18	0 1 18	0 1 18
6	0 1 29	0 1 29	0 1 29	0 1 29	0 1 29
7	0 2 0	0 2 0	0 2 1	0 2 1	0 2 1
8	0 2 12	0 2 12	0 2 12	0 2 12	0 2 13
9	0 2 23	0 2 23	0 2 24	0 2 24	0 2 24
10	0 2 34	0 2 35	0 2 35	0 2 35	0 2 36
20	1 1 29	1 1 29	1 1 30	1 1 31	1 1 31
30	2 0 23	2 0 24	2 0 25	2 0 26	2 0 27
40	2 3 18	2 3 19	2 3 20	2 3 21	2 3 23
50	3 2 12	3 2 14	3 2 15	3 2 17	3 2 19
60	4 1 6	4 1 8	4 1 10	4 1 12	4 1 14
70	5 0 1	5 0 3	5 0 5	5 0 8	5 0 10
80	5 2 35	5 2 38	5 3 0	5 3 3	5 3 6
90	6 1 29	6 1 32	6 1 35	6 1 38	6 2 1
100	7 0 24	7 0 27	7 0 30	7 0 34	7 0 37
200	14 1 8	14 1 14	14 1 21	14 1 27	14 1 34
300	21 1 31	21 2 1	21 2 11	21 2 21	21 2 31
400	28 2 15	28 2 28	28 3 2	28 3 15	28 3 28
500	35 2 39	35 3 16	35 3 32	36 0 9	36 0 25

From 351 to 355 Yards Long.

Width.	351 Yards Long.	352 Yards Long.	353 Yards Long.	354 Yards Long.	355 Yards Long.
Yards.	*A. R. P.*	*A. R. P.*	*A. R. P.*	*A. R. P.*	*A. R. P.*
1	0 0 12	0 0 12	0 0 12	0 0 12	0 0 12
2	0 0 23	0 0 23	0 0 23	0 0 23	0 0 23
3	0 0 35	0 0 35	0 0 35	0 0 35	0 0 35
4	0 1 6	0 1 7	0 1 7	0 1 7	0 1 7
5	0 1 18	0 1 18	0 1 18	0 1 19	0 1 19
6	0 1 30	0 1 30	0 1 30	0 1 30	0 1 30
7	0 2 1	0 2 1	0 2 2	0 2 2	0 2 2
8	0 2 13	0 2 13	0 2 13	0 2 14	0 2 14
9	0 2 24	0 2 25	0 2 25	0 2 25	0 2 26
10	0 2 36	0 2 36	0 2 37	0 2 37	0 2 37
20	1 1 32	1 1 33	1 1 33	1 1 34	1 1 35
30	2 0 28	2 0 29	2 0 30	2 0 31	2 0 32
40	2 3 24	2 3 25	2 3 27	2 3 28	2 3 29
50	3 2 20	3 2 22	3 2 23	3 2 25	3 2 27
60	4 1 16	4 1 18	4 1 20	4 1 22	4 1 24
70	5 0 12	5 0 15	5 0 17	5 0 19	5 0 21
80	5 3 8	5 3 11	5 3 14	5 3 16	5 3 19
90	6 2 4	6 2 7	6 2 10	6 2 13	6 2 16
100	7 1 0	7 1 4	7 1 7	7 1 10	7 1 14
200	14 2 1	14 2 7	14 2 14	14 2 20	14 2 27
300	21 3 1	21 3 11	21 3 21	21 3 31	22 0 1
400	29 0 1	29 0 15	29 0 28	29 1 1	29 1 14
500	36 1 2	36 1 18	36 1 35	36 2 11	36 2 28

From 356 to 360 Yards Long.

Width.	356 Yards Long.	357 Yards Long.	358 Yards Long.	359 Yards Long.	360 Yards Long.
Yards.	*A. R. P.*	*A. R. P.*	*A. R. P.*	*A. R. P.*	*A. R. P.*
1	0 0 12	0 0 12	0 0 12	0 0 12	0 0 12
2	0 0 24	0 0 24	0 0 24	0 0 24	0 0 24
3	0 0 35	0 0 35	0 0 36	0 0 36	0 0 36
4	0 1 7	0 1 7	0 1 7	0 1 7	0 1 8
5	0 1 19	0 1 19	0 1 19	0 1 19	0 1 20
6	0 1 31	0 1 31	0 1 31	0 1 31	0 1 31
7	0 2 2	0 2 3	0 2 3	0 2 3	0 2 3
8	0 2 14	0 2 14	0 2 15	0 2 15	0 2 15
9	0 2 26	0 2 26	0 2 27	0 2 27	0 2 27
10	0 2 38	0 2 38	0 2 38	0 2 39	0 2 39
20	1 1 35	1 1 36	1 1 37	1 1 37	1 1 38
30	2 0 33	2 0 34	2 0 35	2 0 36	2 0 37
40	2 3 31	2 3 32	2 3 33	2 3 35	2 3 36
50	3 2 28	3 2 30	3 2 32	3 2 33	3 2 35
60	4 1 26	4 1 28	4 1 30	4 1 32	4 1 34
70	5 0 24	5 0 26	5 0 28	5 0 31	5 0 33
80	5 3 21	5 3 24	5 3 27	5 3 29	5 3 32
90	6 2 19	6 2 22	6 2 25	6 2 28	6 2 31
100	7 1 17	7 1 20	7 1 23	7 1 27	7 1 30
200	14 2 34	14 3 0	14 3 7	14 3 14	14 3 20
300	22 0 11	22 0 20	22 0 30	22 1 0	22 1 10
400	29 1 27	29 2 1	29 2 14	29 2 27	29 3 0
500	36 3 4	36 3 21	36 3 37	37 0 14	37 0 30

From 361 to 365 Yards Long.

Width.	361 Yards Long.	362 Yards Long.	363 Yards Long.	364 Yards Long.	365 Yards Long.
Yards.	*A. R. P.*	*A. R. P.*	*A. R. P.*	*A. R. P.*	*A. R. P.*
1	0 0 12	0 0 12	0 0 12	0 0 12	0 0 12
2	0 0 24	0 0 24	0 0 24	0 0 24	0 0 24
3	0 0 36	0 0 36	0 0 36	0 0 36	0 0 36
4	0 1 8	0 1 8	0 1 8	0 1 8	0 1 8
5	0 1 20	0 1 20	0 1 20	0 1 20	0 1 20
6	0 1 32	0 1 32	0 1 32	0 1 32	0 1 32
7	0 2 4	0 2 4	0 2 4	0 2 4	0 2 4
8	0 2 15	0 2 16	0 2 16	0 2 16	0 2 17
9	0 2 27	0 2 28	0 2 28	0 2 28	0 2 29
10	0 2 39	0 3 0	0 3 0	0 3 0	0 3 1
20	1 1 39	1 1 39	1 2 0	1 2 1	1 2 1
30	2 0 38	2 0 39	2 1 0	2 1 1	2 1 2
40	2 3 37	2 3 39	3 0 0	3 0 1	3 0 3
50	3 2 37	3 2 38	3 3 0	3 3 2	3 3 3
60	4 1 36	4 1 38	4 2 0	4 2 2	4 2 4
70	5 0 35	5 0 38	5 1 0	5 1 2	5 1 5
80	5 3 35	5 3 37	6 0 0	6 0 3	6 0 5
90	6 2 34	6 2 37	6 3 0	6 3 3	6 3 6
100	7 1 33	7 1 37	7 2 0	7 2 3	7 2 7
200	14 3 27	14 3 33	15 0 0	15 0 7	15 0 13
300	22 1 20	22 1 30	22 2 0	22 2 10	22 2 20
400	29 3 14	29 3 27	30 0 0	30 0 13	30 0 26
500	37 1 7	37 1 23	37 2 0	37 2 17	37 2 33

From 366 to 370 Yards Long.

Width.	366 Yards Long.	367 Yards Long.	368 Yards Long.	369 Yards Long.	370 Yards Long.
Yards.	*A. R. P.*	*A. R. P.*	*A. R. P.*	*A. R. P.*	*A. R. P.*
1	0 0 12	0 0 12	0 0 12	0 0 12	0 0 12
2	0 0 24	0 0 24	0 0 24	0 0 24	0 0 24
3	0 0 36	0 0 36	0 0 36	0 0 37	0 0 37
4	0 1 8	0 1 9	0 1 9	0 1 9	0 1 9
5	0 1 20	0 1 21	0 1 21	0 1 21	0 1 21
6	0 1 33	0 1 33	0 1 33	0 1 33	0 1 33
7	0 2 5	0 2 5	0 2 5	0 2 5	0 2 6
8	0 2 17	0 2 17	0 2 17	0 2 18	0 2 18
9	0 2 29	0 2 29	0 2 29	0 2 30	0 2 30
10	0 3 1	0 3 1	0 3 2	0 3 2	0 3 2
20	1 2 2	1 2 3	1 2 3	1 2 4	1 2 5
30	2 1 3	2 1 4	2 1 5	2 1 6	2 1 7
40	3 0 4	3 0 5	3 0 7	3 0 8	3 0 9
50	3 3 5	3 3 7	3 3 8	3 3 10	3 3 12
60	4 2 6	4 2 8	4 2 10	4 2 12	4 2 14
70	5 1 7	5 1 9	5 1 12	5 1 14	5 1 16
80	6 0 8	6 0 11	6 0 13	6 0 16	6 0 19
90	6 3 9	6 3 12	6 3 15	6 3 18	6 3 21
100	7 2 10	7 2 13	7 2 17	7 2 20	7 2 23
200	15 0 20	15 0 26	15 0 33	15 1 0	15 1 6
300	22 2 30	22 3 0	22 3 10	22 3 20	22 3 29
400	30 1 0	30 1 13	30 1 26	30 1 39	30 2 13
500	37 3 10	37 3 26	38 0 3	38 0 19	38 0 36

From 371 to 375 Yards Long.

Width.	371 Yards Long.	372 Yards Long.	373 Yards Long.	374 Yards Long.	375 Yards Long.
Yards.	*A. R. P.*	*A. R. P.*	*A. R. P.*	*A. R. P.*	*A. R. P.*
1	0 0 12	0 0 12	0 0 12	0 0 12	0 0 12
2	0 0 25	0 0 25	0 0 25	0 0 25	0 0 25
3	0 0 37	0 0 37	0 0 37	0 0 37	0 0 37
4	0 1 9	0 1 9	0 1 9	0 1 9	0 1 10
5	0 1 21	0 1 21	0 1 22	0 1 22	0 1 22
6	0 1 34	0 1 34	0 1 34	0 1 34	0 1 34
7	0 2 6	0 2 6	0 2 6	0 2 7	0 2 7
8	0 2 18	0 2 18	0 2 19	0 2 19	0 2 19
9	0 2 30	0 2 31	0 2 31	0 2 31	0 2 32
10	0 3 3	0 3 3	0 3 3	0 3 4	0 3 4
20	1 2 5	1 2 6	1 2 7	1 2 7	1 2 8
30	2 1 8	2 1 9	2 1 10	2 1 11	2 1 12
40	3 0 11	3 0 12	3 0 13	3 0 15	3 0 16
50	3 3 13	3 3 15	3 3 17	3 3 18	3 3 20
60	4 2 16	4 2 18	4 2 20	4 2 22	4 2 24
70	5 1 19	5 1 21	5 1 23	5 1 25	5 1 28
80	6 0 21	6 0 24	6 0 26	6 0 29	6 0 32
90	6 3 24	6 3 27	6 3 30	6 3 33	6 3 36
100	7 2 26	7 2 30	7 2 33	7 2 36	7 3 0
200	15 1 13	15 1 20	15 1 26	15 1 33	15 1 39
300	22 3 39	23 0 9	23 0 19	23 0 29	23 0 39
400	30 2 26	30 2 39	30 3 12	30 3 25	30 3 39
500	38 1 12	38 1 29	38 2 5	38 2 22	38 2 38

From 376 to 380 Yards Long.

Width.	376 Yards Long.	377 Yards Long.	378 Yards Long.	379 Yards Long.	380 Yards Long.
Yards.	A. R. P.	A. R. P.	A. R. P.	A. R. P.	A. R. P.
1	0 0 12	0 0 12	0 0 12	0 0 13	0 0 13
2	0 0 25	0 0 25	0 0 25	0 0 25	0 0 25
3	0 0 37	0 0 37	0 0 37	0 0 38	0 0 38
4	0 1 10	0 1 10	0 1 10	0 1 10	0 1 10
5	0 1 22	0 1 22	0 1 22	0 1 23	0 1 23
6	0 1 35	0 1 35	0 1 35	0 1 35	0 1 35
7	0 2 7	0 2 7	0 2 7	0 2 8	0 2 8
8	0 2 19	0 2 20	0 2 20	0 2 20	0 2 20
9	0 2 32	0 2 32	0 2 32	0 2 33	0 2 33
10	0 3 4	0 3 5	0 3 5	0 3 5	0 3 6
20	1 2 9	1 2 9	1 2 10	1 2 11	1 2 11
30	2 1 13	2 1 14	2 1 15	2 1 16	2 1 17
40	3 0 17	3 0 19	3 0 20	3 0 21	3 0 22
50	3 3 21	3 3 23	3 3 25	3 3 26	3 3 28
60	4 2 26	4 2 28	4 2 30	4 2 32	4 2 34
70	5 1 30	5 1 32	5 1 35	5 1 37	5 1 39
80	6 0 34	6 0 37	6 1 0	6 1 2	6 1 5
90	6 3 39	7 0 2	7 0 5	7 0 8	7 0 11
100	7 3 3	7 3 6	7 3 10	7 3 13	7 3 16
200	15 2 6	15 2 13	15 2 19	15 2 26	15 2 32
300	23 1 9	23 1 19	23 1 29	23 1 39	23 2 9
400	31 0 12	31 0 25	31 0 38	31 1 12	31 1 25
500	38 3 15	38 3 31	39 0 8	39 0 24	39 1 1

From 381 to 385 Yards Long.

Width.	381 Yards Long.	382 Yards Long.	383 Yards Long.	384 Yards Long.	385 Yards Long.
Yards.	*A. R. P.*	*A. R. P.*	*A. R. P.*	*A. R. P.*	*A. R. P.*
1	0 0 13	0 0 13	0 0 13	0 0 13	0 0 13
2	0 0 25	0 0 25	0 0 25	0 0 25	0 0 25
3	0 0 38	0 0 38	0 0 38	0 0 38	0 0 38
4	0 1 10	0 1 11	0 1 11	0 1 11	0 1 11
5	0 1 23	0 1 23	0 1 23	0 1 23	0 1 24
6	0 1 36	0 1 36	0 1 36	0 1 36	0 1 36
7	0 2 8	0 2 8	0 2 9	0 2 9	0 2 9
8	0 2 21	0 2 21	0 2 21	0 2 22	0 2 22
9	0 2 33	0 2 34	0 2 34	0 2 34	0 2 35
10	0 3 6	0 3 6	0 3 7	0 3 7	0 3 7
20	1 2 12	1 2 13	1 2 13	1 2 14	1 2 15
30	2 1 18	2 1 19	2 1 20	2 1 21	2 1 22
40	3 0 24	3 0 25	3 0 26	3 0 28	3 0 29
50	3 3 30	3 3 31	3 3 33	3 3 35	3 3 36
60	4 2 36	4 2 38	4 3 0	4 3 2	4 3 4
70	5 2 2	5 2 4	5 2 6	5 2 9	5 2 11
80	6 1 8	6 1 10	6 1 13	6 1 16	6 1 18
90	7 0 14	7 0 17	7 0 20	7 0 22	7 0 25
100	7 3 20	7 3 23	7 3 26	7 3 29	7 3 33
200	15 2 39	15 3 6	15 3 12	15 3 19	15 3 25
300	23 2 19	23 2 28	23 2 38	23 3 8	23 3 18
400	31 1 38	31 2 11	31 2 24	31 2 38	31 3 11
500	39 1 18	39 1 34	39 2 11	39 2 27	39 3 4

From 386 to 390 Yards Long.

Width.	386 Yards Long.	387 Yards Long.	388 Yards Long.	389 Yards Long.	390 Yards Long.
Yards.	*A. R. P.*	*A. R. P.*	*A. R. P.*	*A. R. P.*	*A. R. P.*
1	0 0 13	0 0 13	0 0 13	0 0 13	0 0 13
2	0 0 26	0 0 26	0 0 26	0 0 26	0 0 26
3	0 0 38	0 0 38	0 0 38	0 0 39	0 0 39
4	0 1 11	0 1 11	0 1 11	0 1 11	0 1 12
5	0 1 24	0 1 24	0 1 24	0 1 24	0 1 24
6	0 1 37	0 1 37	0 1 37	0 1 37	0 1 37
7	0 2 9	0 2 10	0 2 10	0 2 10	0 2 10
8	0 2 22	0 2 22	0 2 23	0 2 23	0 2 23
9	0 2 35	0 2 35	0 2 35	0 2 36	0 2 36
10	0 3 8	0 3 8	0 3 8	0 3 9	0 3 9
20	1 2 15	1 2 16	1 2 17	1 2 17	1 2 18
30	2 1 23	2 1 24	2 1 25	2 1 26	2 1 27
40	3 0 30	3 0 32	3 0 33	3 0 34	3 0 36
50	3 3 38	4 0 0	4 0 1	4 0 3	4 0 5
60	4 3 6	4 3 8	4 3 10	4 3 12	4 3 14
70	5 2 13	5 2 16	5 2 18	5 2 20	5 2 22
80	6 1 21	6 1 23	6 1 26	6 1 29	6 1 31
90	7 0 28	7 0 31	7 0 34	7 0 37	7 1 0
100	7 3 36	7 3 39	8 0 3	8 0 6	8 0 9
200	15 3 32	15 3 39	16 0 5	16 0 12	16 0 19
300	23 3 28	23 3 38	24 0 8	24 0 18	24 0 28
400	31 3 24	31 3 37	32 0 11	32 0 24	32 0 37
500	39 3 20	39 3 37	40 0 13	40 0 30	40 1 6

From 391 to 395 Yards Long.

Width.	391 Yards Long.	392 Yards Long.	393 Yards Long.	394 Yards Long.	395 Yards Long.
Yards.	*A. R. P.*	*A. R. P.*	*A. R. P.*	*A. R. P.*	*A. R. P.*
1	0 0 13	0 0 13	0 0 13	0 0 13	0 0 13
2	0 0 26	0 0 26	0 0 26	0 0 26	0 0 26
3	0 0 39	0 0 39	0 0 39	0 0 39	0 0 39
4	0 1 12	0 1 12	0 1 12	0 1 12	0 1 12
5	0 1 25	0 1 25	0 1 25	0 1 25	0 1 25
6	0 1 38	0 1 38	0 1 38	0 1 38	0 1 38
7	0 2 10	0 2 11	0 2 11	0 2 11	0 2 11
8	0 2 23	0 2 24	0 2 24	0 2 24	0 2 24
9	0 2 36	0 2 37	0 2 37	0 2 37	0 2 38
10	0 3 9	0 3 10	0 3 10	0 3 10	0 3 11
20	1 2 19	1 2 19	1 2 20	1 2 20	1 2 21
30	2 1 28	2 1 29	2 1 30	2 1 31	2 1 32
40	3 0 37	3 0 38	3 1 0	3 1 1	3 1 2
50	4 0 6	4 0 8	4 0 10	4 0 11	4 0 13
60	4 3 16	4 3 18	4 3 20	4 3 21	4 3 23
70	5 2 25	5 2 27	5 2 29	5 2 32	5 2 34
80	6 1 34	6 1 37	6 1 39	6 2 2	6 2 5
90	7 1 3	7 1 6	7 1 9	7 1 12	7 1 15
100	8 0 13	8 0 16	8 0 19	8 0 22	8 0 26
200	16 0 25	16 0 32	16 0 38	16 1 5	16 1 12
300	24 0 38	24 1 8	24 1 18	24 1 27	24 1 37
400	32 1 10	32 1 23	32 1 37	32 2 10	32 2 23
500	40 1 23	40 1 39	40 2 16	40 2 32	40 3 9

From 396 to 400 Yards Long.

Width.	396 Yards Long.	397 Yards Long.	398 Yards Long.	399 Yards Long.	400 Yards Long.
Yards.	*A. R. P.*	*A. R. P.*	*A. R. P.*	*A. R. P.*	*A. R. P.*
1	0 0 13	0 0 13	0 0 13	0 0 13	0 0 13
2	0 0 26	0 0 26	0 0 26	0 0 26	0 0 26
3	0 0 39	0 0 39	0 0 39	0 1 0	0 1 0
4	0 1 12	0 1 12	0 1 13	0 1 13	0 1 13
5	0 1 25	0 1 26	0 1 26	0 1 26	0 1 26
6	0 1 39	0 1 39	0 1 39	0 1 39	0 1 39
7	0 2 12	0 2 12	0 2 12	0 2 12	0 2 13
8	0 2 25	0 2 25	0 2 25	0 2 26	0 2 26
9	0 2 38	0 2 38	0 2 38	0 2 39	0 2 39
10	0 3 11	0 3 11	0 3 12	0 3 12	0 3 12
20	1 2 22	1 2 22	1 2 23	1 2 24	1 2 24
30	2 1 33	2 1 34	2 1 35	2 1 36	2 1 37
40	3 1 4	3 1 5	3 1 6	3 1 8	3 1 9
50	4 0 15	4 0 16	4 0 18	4 0 20	4 0 21
60	4 3 25	4 3 27	4 3 29	4 3 31	4 3 33
70	5 2 36	5 2 39	5 3 1	5 3 3	5 3 6
80	6 2 7	6 2 10	6 2 13	6 2 15	6 2 18
90	7 1 18	7 1 21	7 1 24	7 1 27	7 1 30
100	8 0 29	8 0 32	8 0 36	8 0 39	8 1 2
200	16 1 18	16 1 25	16 1 31	16 1 38	16 2 5
300	24 2 7	24 2 17	24 2 27	24 2 37	24 3 7
400	32 2 36	32 3 10	32 3 23	32 3 36	33 0 9
500	40 3 25	41 0 2	41 0 19	41 0 35	41 1 12

9*

From 401 to 405 Yards Long.

Width.	401 Yards Long.	402 Yards Long.	403 Yards Long.	404 Yards Long.	405 Yards Long.
Yards.	*A. R. P.*	*A. R. P.*	*A. R. P.*	*A. R. P.*	*A. R. P.*
1	0 0 13	0 0 13	0 0 13	0 0 13	0 0 13
2	0 0 27	0 0 27	0 0 27	0 0 27	0 0 27
3	0 1 0	0 1 0	0 1 0	0 1 0	0 1 0
4	0 1 13	0 1 13	0 1 13	0 1 13	0 1 14
5	0 1 26	0 1 26	0 1 27	0 1 27	0 1 27
6	0 2 0	0 2 0	0 2 0	0 2 0	0 2 0
7	0 2 13	0 2 13	0 2 13	0 2 13	0 2 14
8	0 2 26	0 2 26	0 2 27	0 2 27	0 2 27
9	0 2 39	0 3 0	0 3 0	0 3 0	0 3 0
10	0 3 13	0 3 13	0 3 13	0 3 14	0 3 14
20	1 2 25	1 2 26	1 2 26	1 2 27	1 2 28
30	2 1 38	2 1 39	2 2 0	2 2 1	2 2 2
40	3 1 10	3 1 12	3 1 13	3 1 14	3 1 16
50	4 0 23	4 0 24	4 0 26	4 0 28	4 0 29
60	4 3 35	4 3 37	4 3 39	5 0 1	5 0 3
70	5 3 8	5 3 10	5 3 13	5 3 15	5 3 17
80	6 2 20	6 2 23	6 2 26	6 2 28	6 2 31
90	7 1 33	7 1 36	7 1 39	7 2 2	7 2 5
100	8 1 6	8 1 9	8 1 12	8 1 16	8 1 19
200	16 2 11	16 2 18	16 2 24	16 2 31	16 2 38
300	24 3 17	24 3 27	24 3 37	25 0 7	25 0 17
400	33 0 22	33 0 36	33 1 9	33 1 22	33 1 35
500	41 1 28	41 2 5	41 2 21	41 2 38	41 3 14

From 406 to 410 Yards Long.

Width.	406 Yards Long.	407 Yards Long.	408 Yards Long.	409 Yards Long.	410 Yards Long.
Yards.	*A. R. P.*	*A. R. P.*	*A. R. P.*	*A. R. P.*	*A. R. P.*
1	0 0 13	0 0 13	0 0 13	0 0 14	0 0 14
2	0 0 27	0 0 27	0 0 27	0 0 27	0 0 27
3	0 1 0	0 1 0	0 1 0	0 1 1	0 1 1
4	0 1 14	0 1 14	0 1 14	0 1 14	0 1 14
5	0 1 27	0 1 27	0 1 27	0 1 28	0 1 28
6	0 2 1	0 2 1	0 2 1	0 2 1	0 2 1
7	0 2 14	0 2 14	0 2 14	0 2 15	0 2 15
8	0 2 27	0 2 28	0 2 28	0 2 28	0 2 28
9	0 3 1	0 3 1	0 3 1	0 3 2	0 3 2
10	0 3 14	0 3 15	0 3 15	0 3 15	0 3 16
20	1 2 28	1 2 29	1 2 30	1 2 30	1 2 31
30	2 2 3	2 2 4	2 2 5	2 2 6	2 2 7
40	3 1 17	3 1 18	3 1 20	3 1 21	3 1 22
50	4 0 31	4 0 33	4 0 34	4 0 36	4 0 38
60	5 0 5	5 0 7	5 0 9	5 0 11	5 0 13
70	5 3 20	5 3 22	5 3 24	5 3 26	5 3 29
80	6 2 34	6 2 36	6 2 39	6 3 2	6 3 4
90	7 2 8	7 2 11	7 2 14	7 2 17	7 2 20
100	8 1 22	8 1 25	8 1 29	8 1 32	8 1 35
200	16 3 4	16 3 11	16 3 18	16 3 24	16 3 31
300	25 0 26	25 0 36	25 1 6	25 1 16	25 1 26
400	33 2 9	33 2 22	33 2 35	33 3 8	33 3 21
500	41 3 31	42 0 7	42 0 24	42 1 0	42 1 17

From 411 to 415 Yards Long.

Width.	411 Yards Long.	412 Yards Long.	413 Yards Long.	414 Yards Long.	415 Yards Long.
Yards.	*A. R. P.*	*A. R. P.*	*A. R. P.*	*A. R. P.*	*A. R. P.*
1	0 0 14	0 0 14	0 0 14	0 0 14	0 0 14
2	0 0 27	0 0 27	0 0 27	0 0 27	0 0 27
3	0 1 1	0 1 1	0 1 1	0 1 1	0 1 1
4	0 1 14	0 1 14	0 1 15	0 1 15	0 1 15
5	0 1 28	0 1 28	0 1 28	0 1 28	0 1 29
6	0 2 2	0 2 2	0 2 2	0 2 2	0 2 2
7	0 2 15	0 2 15	0 2 16	0 2 16	0 2 16
8	0 2 29	0 2 29	0 2 29	0 2 29	0 2 30
9	0 3 2	0 3 3	0 3 3	0 3 3	0 3 3
10	0 3 16	0 3 16	0 3 17	0 3 17	0 3 17
20	1 2 32	1 2 32	1 2 33	1 2 34	1 2 34
30	2 2 8	2 2 9	2 2 10	2 2 11	2 2 12
40	3 1 23	3 1 25	3 1 26	3 1 27	3 1 29
50	4 0 39	4 1 1	4 1 3	4 1 4	4 1 6
60	5 0 15	5 0 17	5 0 19	5 0 21	5 0 23
70	5 3 31	5 3 33	5 3 36	5 3 38	6 0 0
80	6 3 7	6 3 10	6 3 12	6 3 15	6 3 18
90	7 2 23	7 2 26	7 2 29	7 2 32	7 2 35
100	8 1 39	8 2 2	8 2 5	8 2 9	8 2 12
200	16 3 37	17 0 4	17 0 11	17 0 17	17 0 24
300	25 1 36	25 2 6	25 2 16	25 2 26	25 2 36
400	33 3 35	34 0 8	34 0 21	34 0 34	34 1 8
500	42 1 33	42 2 10	42 2 26	42 3 3	42 3 20

From 416 to 420 Yards Long.

Width.	416 Yards Long.	417 Yards Long.	418 Yards Long.	419 Yards Long.	420 Yards Long.
Yards.	*A. R. P.*	*A. R. P.*	*A. R. P.*	*A. R. P.*	*A. R. P.*
1	0 0 14	0 0 14	0 0 14	0 0 14	0 0 14
2	0 0 28	0 0 28	0 0 28	0 0 28	0 0 28
3	0 1 1	0 1 1	0 1 1	0 1 2	0 1 2
4	0 1 15	0 1 15	0 1 15	0 1 15	0 1 16
5	0 1 29	0 1 29	0 1 29	0 1 29	0 1 29
6	0 2 3	0 2 3	0 2 3	0 2 3	0 2 3
7	0 2 16	0 2 16	0 2 17	0 2 17	0 2 17
8	0 2 30	0 2 30	0 2 31	0 2 31	0 2 31
9	0 3 4	0 3 4	0 3 4	0 3 5	0 3 5
10	0 3 18	0 3 18	0 3 18	0 3 19	0 3 19
20	1 2 35	1 2 36	1 2 36	1 2 37	1 2 38
30	2 2 13	2 2 14	2 2 15	2 2 16	2 2 17
40	3 1 30	3 1 31	3 1 33	3 1 34	3 1 35
50	4 1 8	4 1 9	4 1 11	4 1 13	4 1 14
60	5 0 25	5 0 27	5 0 29	5 0 31	5 0 33
70	6 0 3	6 0 5	6 0 7	6 0 10	6 0 12
80	6 3 20	6 3 23	6 3 25	6 3 28	6 3 31
90	7 2 38	7 3 1	7 3 4	7 3 7	7 3 10
100	8 2 15	8 2 19	8 2 22	8 2 25	8 2 28
200	17 0 30	17 0 37	17 1 4	17 1 10	17 1 17
300	25 3 6	25 3 16	25 3 25	25 3 35	26 0 5
400	34 1 21	34 1 34	34 2 7	34 2 20	34 2 34
500	42 3 36	43 0 13	43 0 29	43 1 6	43 1 22

From 421 to 425 Yards Long.

Width.	421 Yards Long.	422 Yards Long.	423 Yards Long.	424 Yards Long.	425 Yards Long.
Yards.	*A. R. P.*	*A. R. P.*	*A. R. P.*	*A. R. P.*	*A. R. P.*
1	0 0 14	0 0 14	0 0 14	0 0 14	0 0 14
2	0 0 28	0 0 28	0 0 28	0 0 28	0 0 28
3	0 1 2	0 1 2	0 1 2	0 1 2	0 1 2
4	0 1 16	0 1 16	0 1 16	0 1 16	0 1 16
5	0 1 30	0 1 30	0 1 30	0 1 30	0 1 30
6	0 2 4	0 2 4	0 2 4	0 2 4	0 2 4
7	0 2 17	0 2 18	0 2 18	0 2 18	0 2 18
8	0 2 31	0 2 32	0 2 32	0 2 32	0 2 32
9	0 3 5	0 3 6	0 3 6	0 3 6	0 3 6
10	0 3 19	0 3 20	0 3 20	0 3 20	0 3 20
20	1 2 38	1 2 39	1 3 0	1 3 0	1 3 1
30	2 2 18	2 2 19	2 2 20	2 2 20	2 2 21
40	3 1 37	3 1 38	3 1 39	3 2 1	3 2 2
50	4 1 16	4 1 18	4 1 19	4 1 21	4 1 22
60	5 0 35	5 0 37	5 0 39	5 1 1	5 1 3
70	6 0 14	6 0 17	6 0 19	6 0 21	6 0 23
80	6 3 33	6 3 36	6 3 39	7 0 1	7 0 4
90	7 3 13	7 3 16	7 3 19	7 3 21	7 3 24
100	8 2 32	8 2 35	8 2 38	8 3 2	8 3 5
200	17 1 23	17 1 30	17 1 37	17 2 3	17 2 10
300	26 0 15	26 0 25	26 0 35	26 1 5	26 1 15
400	34 3 7	34 3 20	34 3 33	35 0 7	35 0 20
500	43 1 39	43 2 15	43 2 32	43 3 8	43 3 25

From 426 to 430 Yards Long.

Width.	426 Yards Long.	427 Yards Long.	428 Yards Long.	429 Yards Long.	430 Yards Long.
Yards.	*A. R. P.*	*A. R. P.*	*A. R. P.*	*A. R. P.*	*A. R. P.*
1	0 0 14	0 0 14	0 0 14	0 0 14	0 0 14
2	0 0 28	0 0 28	0 0 28	0 0 28	0 0 28
3	0 1 2	0 1 2	0 1 2	0 1 3	0 1 3
4	0 1 16	0 1 16	0 1 17	0 1 17	0 1 17
5	0 1 30	0 1 31	0 1 31	0 1 31	0 1 31
6	0 2 4	0 2 5	0 2 5	0 2 5	0 2 5
7	0 2 19	0 2 19	0 2 19	0 2 19	0 2 20
8	0 2 33	0 2 33	0 2 33	0 2 33	0 2 34
9	0 3 7	0 3 7	0 3 7	0 3 8	0 3 8
10	0 3 21	0 3 21	0 3 21	0 3 22	0 3 22
20	1 3 2	1 3 2	1 3 3	1 3 4	1 3 4
30	2 2 22	2 2 23	2 2 24	2 2 25	2 2 26
40	3 2 3	3 2 5	3 2 6	3 2 7	3 2 9
50	4 1 24	4 1 26	4 1 27	4 1 29	4 1 31
60	5 1 5	5 1 7	5 1 9	5 1 11	5 1 13
70	6 0 26	6 0 28	6 0 30	6 0 33	6 0 35
80	7 0 7	7 0 9	7 0 12	7 0 15	7 0 17
90	7 3 27	7 3 30	7 3 33	7 3 36	7 3 39
100	8 3 8	8 3 12	8 3 15	8 3 18	8 3 21
200	17 2 17	17 2 23	17 2 30	17 2 36	17 3 3
300	26 1 25	26 1 35	26 2 5	26 2 15	26 2 24
400	35 0 33	35 1 6	35 1 20	35 1 33	35 2 6
500	44 0 1	44 0 18	44 0 34	44 1 11	44 1 27

From 431 to 435 Yards Long.

Width.	431 Yards Long.	432 Yards Long.	433 Yards Long.	434 Yards Long.	435 Yards Long.
Yards.	*A. R. P.*	*A. R. P.*	*A. R. P.*	*A. R. P.*	*A. R. P.*
1	0 0 14	0 0 14	0 0 14	0 0 14	0 0 14
2	0 0 28	0 0 29	0 0 29	0 0 29	0 0 29
3	0 1 3	0 1 3	0 1 3	0 1 3	0 1 3
4	0 1 17	0 1 17	0 1 17	0 1 17	0 1 18
5	0 1 31	0 1 31	0 1 32	0 1 32	0 1 32
6	0 2 5	0 2 6	0 2 6	0 2 6	0 2 6
7	0 2 20	0 2 20	0 2 20	0 2 20	0 2 21
8	0 2 34	0 2 34	0 2 35	0 2 35	0 2 35
9	0 3 8	0 3 9	0 3 9	0 3 9	0 3 9
10	0 3 22	0 3 23	0 3 23	0 3 23	0 3 24
20	1 3 5	1 3 6	1 3 6	1 3 7	1 3 8
30	2 2 27	2 2 28	2 2 29	2 2 30	2 2 31
40	3 2 10	3 2 11	3 2 13	3 2 14	3 2 15
50	4 1 32	4 1 34	4 1 36	4 1 37	4 1 39
60	5 1 15	5 1 17	5 1 19	5 1 21	5 1 23
70	6 0 37	6 1 0	6 1 2	6 1 4	6 1 7
80	7 0 20	7 0 22	7 0 25	7 0 28	7 0 30
90	8 0 2	8 0 5	8 0 8	8 0 11	8 0 14
100	8 3 25	8 3 28	8 3 31	8 3 35	8 3 38
200	17 3 10	17 3 16	17 3 23	17 3 29	17 3 36
300	26 2 34	26 3 4	26 3 14	26 3 24	26 3 34
400	35 2 19	35 2 32	35 3 6	35 3 19	35 3 32
500	44 2 4	44 2 20	44 2 37	44 3 14	44 3 30

From 436 to 440 Yards Long.

Width.	436 Yards Long.	437 Yards Long.	438 Yards Long.	439 Yards Long.	440 Yards Long.
Yards.	*A. R. P.*	*A. R. P.*	*A. R. P.*	*A. R. P.*	*A. R. P.*
1	0 0 14	0 0 14	0 0 14	0 0 15	0 0 15
2	0 0 29	0 0 29	0 0 29	0 0 29	0 0 29
3	0 1 3	0 1 3	0 1 3	0 1 4	0 1 4
4	0 1 18	0 1 18	0 1 18	0 1 18	0 1 18
5	0 1 32	0 1 32	0 1 32	0 1 33	0 1 33
6	0 2 6	0 2 7	0 2 7	0 2 7	0 2 7
7	0 2 21	0 2 21	0 2 21	0 2 22	0 2 22
8	0 2 35	0 2 36	0 2 36	0 2 36	0 2 36
9	0 3 10	0 3 10	0 3 10	0 3 11	0 3 11
10	0 3 24	0 3 24	0 3 25	0 3 25	0 3 25
20	1 3 8	1 3 9	1 3 10	1 3 10	1 3 11
30	2 2 32	2 2 33	2 2 34	2 2 35	2 2 36
40	3 2 17	3 2 18	3 2 19	3 2 20	3 2 22
50	4 2 1	4 2 2	4 2 4	4 2 6	4 2 7
60	5 1 25	5 1 27	5 1 29	5 1 31	5 1 33
70	6 1 9	6 1 11	6 1 14	6 1 16	6 1 18
80	7 0 33	7 0 36	7 0 38	7 1 1	7 1 4
90	8 0 17	8 0 20	8 0 23	8 0 26	8 0 29
100	9 0 1	9 0 5	9 0 8	9 0 11	9 0 15
200	18 0 3	18 0 9	18 0 16	18 0 22	18 0 29
300	27 0 4	27 0 14	27 0 24	27 0 34	27 1 4
400	36 0 5	36 0 19	36 0 32	36 1 5	36 1 18
500	45 0 7	45 0 23	45 1 0	45 1 16	45 1 33

From 441 to 445 Yards Long.

Width.	441 Yards Long.	442 Yards Long.	443 Yards Long.	444 Yards Long.	445 Yards Long.
Yards.	*A. R. P.*	*A. R. P.*	*A. R. P.*	*A. R. P.*	*A. R. P.*
1	0 0 15	0 0 15	0 0 15	0 0 15	0 0 15
2	0 0 29	0 0 29	0 0 29	0 0 29	0 0 29
3	0 1 4	0 1 4	0 1 4	0 1 4	0 1 4
4	0 1 18	0 1 18	0 1 19	0 1 19	0 1 19
5	0 1 33	0 1 33	0 1 33	0 1 33	0 1 34
6	0 2 7	0 2 8	0 2 8	0 2 8	0 2 8
7	0 2 22	0 2 22	0 2 23	0 2 23	0 2 23
8	0 2 37	0 2 37	0 2 37	0 2 37	0 2 38
9	0 3 11	0 3 12	0 3 12	0 3 12	0 3 12
10	0 3 26	0 3 26	0 3 26	0 3 27	0 3 27
20	1 3 12	1 3 12	1 3 13	1 3 14	1 3 14
30	2 2 37	2 2 38	2 2 39	2 3 0	2 3 1
40	3 2 23	3 2 24	3 2 26	3 2 27	3 2 28
50	4 2 9	4 2 11	4 2 12	4 2 14	4 2 16
60	5 1 35	5 1 37	5 1 39	5 2 1	5 2 3
70	6 1 20	6 1 23	6 1 25	6 1 27	6 1 30
80	7 1 6	7 1 9	7 1 12	7 1 14	7 1 17
90	8 0 32	8 0 35	8 0 38	8 1 1	8 1 4
100	9 0 18	9 0 21	9 0 24	9 0 28	9 0 31
200	18 0 36	18 1 2	18 1 9	18 1 16	18 1 22
300	27 1 14	27 1 23	27 1 33	27 2 3	27 2 13
400	36 1 31	36 2 5	36 2 18	36 2 31	36 3 4
500	45 2 9	45 2 26	45 3 2	45 3 19	45 3 35

From 446 to 450 Yards Long.

Width.	446 Yards Long.	447 Yards Long.	448 Yards Long.	449 Yards Long.	450 Yards Long.
Yards.	*A. R. P.*	*A. R. P.*	*A. R. P.*	*A. R. P.*	*A. R. P.*
1	0 0 15	0 0 15	0 0 15	0 0 15	0 0 15
2	0 0 29	0 0 30	0 0 30	0 0 30	0 0 30
3	0 1 4	0 1 4	0 1 4	0 1 5	0 1 5
4	0 1 19	0 1 19	0 1 19	0 1 19	0 1 20
5	0 1 34	0 1 34	0 1 34	0 1 34	0 1 34
6	0 2 8	0 2 9	0 2 9	0 2 9	0 2 9
7	0 2 23	0 2 23	0 2 24	0 2 24	0 2 24
8	0 2 38	0 2 38	0 2 38	0 2 39	0 2 39
9	0 3 13	0 3 13	0 3 13	0 3 14	0 3 14
10	0 3 27	0 3 28	0 3 28	0 3 28	0 3 29
20	1 3 15	1 3 16	1 3 16	1 3 17	1 3 18
30	2 3 2	2 3 3	2 3 4	2 3 5	2 3 6
40	3 2 30	3 2 31	3 2 32	3 2 34	3 2 35
50	4 2 17	4 2 19	4 2 20	4 2 22	4 2 24
60	5 2 5	5 2 7	5 2 9	5 2 11	5 2 13
70	6 1 32	6 1 34	6 1 37	6 1 39	6 2 1
80	7 1 20	7 1 22	7 1 25	7 1 27	7 1 30
90	8 1 7	8 1 10	8 1 13	8 1 16	8 1 19
100	9 0 34	9 0 38	9 1 1	9 1 4	9 1 8
200	18 1 29	18 1 35	18 2 2	18 2 9	18 2 15
300	27 2 23	27 2 33	27 3 3	27 3 13	27 3 23
400	36 3 18	36 3 31	37 0 4	37 0 17	37 0 30
500	46 0 12	46 0 28	46 1 5	46 1 21	46 1 38

From 451 to 455 Yards Long.

Width.	451 Yards Long.	452 Yards Long.	453 Yards Long.	454 Yards Long.	455 Yards Long.
Yards.	*A. R. P.*	*A. R. P.*	*A. R. P.*	*A. R. P.*	*A. R. P.*
1	0 0 15	0 0 15	0 0 15	0 0 15	0 0 15
2	0 0 30	0 0 30	0 0 30	0 0 30	0 0 30
3	0 1 5	0 1 5	0 1 5	0 1 5	0 1 5
4	0 1 20	0 1 20	0 1 20	0 1 20	0 1 20
5	0 1 35	0 1 35	0 1 35	0 1 35	0 1 35
6	0 2 9	0 2 10	0 2 10	0 2 10	0 2 10
7	0 2 24	0 2 25	0 2 25	0 2 25	0 2 25
8	0 2 39	0 3 0	0 3 0	0 3 0	0 3 0
9	0 3 14	0 3 14	0 3 15	0 3 15	0 3 15
10	0 3 29	0 3 29	0 3 30	0 3 30	0 3 30
20	1 3 18	1 3 19	1 3 20	1 3 20	1 3 21
30	2 3 7	2 3 8	2 3 9	2 3 10	2 3 11
40	3 2 36	3 2 38	3 2 39	3 3 0	3 3 2
50	4 2 25	4 2 27	4 2 29	4 2 30	4 2 32
60	5 2 15	5 2 17	5 2 19	5 2 20	5 2 22
70	6 2 4	6 2 6	6 2 8	6 2 11	6 2 13
80	7 1 33	7 1 35	7 1 38	7 2 1	7 2 3
90	8 1 22	8 1 25	8 1 28	8 1 31	8 1 34
100	9 1 11	9 1 14	9 1 18	9 1 21	9 1 24
200	18 2 12	18 2 28	18 2 35	18 3 2	18 3 8
300	27 3 33	28 0 8	28 0 13	28 0 22	28 0 32
400	37 1 4	37 1 17	37 1 30	37 2 3	37 2 17
500	46 2 15	46 2 31	46 3 8	46 3 24	47 0 1

From 456 to 460 Yards Long.

Width.	456 Yards Long.	457 Yards Long.	458 Yards Long.	459 Yards Long.	460 Yards Long.
Yards.	*A. R. P.*	*A. R. P.*	*A. R. P.*	*A. R. P.*	*A. R. P.*
1	0 0 15	0 0 15	0 0 15	0 0 15	0 0 15
2	0 0 30	0 0 30	0 0 30	0 0 30	0 0 30
3	0 1 5	0 1 5	0 1 5	0 1 6	0 1 6
4	0 1 20	0 1 20	0 1 21	0 1 21	0 1 21
5	0 1 35	0 1 36	0 1 36	0 1 36	0 1 36
6	0 2 10	0 2 11	0 2 11	0 2 11	0 2 11
7	0 2 26	0 2 26	0 2 26	0 2 26	0 2 26
8	0 3 1	0 3 1	0 3 1	0 3 1	0 3 2
9	0 3 16	0 3 16	0 3 16	0 3 17	0 3 17
10	0 3 31	0 3 31	0 3 31	0 3 32	0 3 32
20	1 3 21	1 3 22	1 3 23	1 3 23	1 3 24
30	2 3 12	2 3 13	2 3 14	2 3 15	2 3 16
40	3 3 3	3 3 4	3 3 6	3 3 7	3 3 8
50	4 2 34	4 2 35	4 2 37	4 2 39	4 2 0
60	5 2 24	5 2 26	5 2 28	5 2 30	5 2 32
70	6 2 15	6 2 18	6 2 20	6 2 22	6 2 24
80	7 2 6	7 2 9	7 2 11	7 2 14	7 2 17
90	8 1 37	8 2 0	8 2 3	8 2 6	8 2 9
100	9 1 27	9 1 31	9 1 34	9 1 37	9 2 1
200	18 3 15	18 3 21	18 3 28	18 3 35	19 0 1
300	28 1 2	28 1 12	28 1 22	28 1 32	28 2 2
400	37 2 30	37 3 3	37 3 16	37 3 29	38 0 3
500	47 0 17	47 0 34	47 1 10	47 1 27	47 2 3

From 461 to 465 Yards Long.

Width.	461 Yards Long.	462 Yards Long.	463 Yards Long.	464 Yards Long.	465 Yards Long.
Yards.	*A. R. P.*	*A. R. P.*	*A. R. P.*	*A. R. P.*	*A. R. P.*
1	0 0 15	0 0 15	0 0 15	0 0 15	0 0 15
2	0 0 30	0 0 31	0 0 31	0 0 31	0 0 31
3	0 1 6	0 1 6	0 1 6	0 1 6	0 1 6
4	0 1 21	0 1 21	0 1 21	0 1 21	0 1 21
5	0 1 36	0 1 36	0 1 37	0 1 37	0 1 37
6	0 2 11	0 2 12	0 2 12	0 2 12	0 2 12
7	0 2 27	0 2 27	0 2 27	0 2 27	0 2 28
8	0 3 2	0 3 2	0 3 2	0 3 3	0 3 3
9	0 3 17	0 3 17	0 3 18	0 3 18	0 3 18
10	0 3 32	0 3 33	0 3 33	0 3 33	0 3 34
20	1 3 25	1 3 25	1 3 26	1 3 27	1 3 27
30	2 3 17	2 3 18	2 3 19	2 3 20	2 3 21
40	3 3 10	3 3 11	3 3 12	3 3 14	3 3 15
50	4 3 2	4 3 4	4 3 5	4 3 7	4 3 9
60	5 2 34	5 2 36	5 2 38	5 3 0	5 3 2
70	6 2 27	6 2 29	6 2 31	6 2 34	6 2 36
80	7 2 19	7 2 22	7 2 24	7 2 27	7 2 30
90	8 2 12	8 2 15	8 2 18	8 2 20	8 2 23
100	9 2 4	9 2 7	9 2 11	9 2 14	9 2 17
200	19 0 8	19 0 15	19 0 21	19 0 28	19 0 34
300	28 2 12	28 2 22	28 2 32	28 3 2	28 3 12
400	38 0 16	38 0 29	38 1 2	38 1 16	38 1 29
500	47 2 20	47 2 36	47 3 13	47 3 29	48 0 6

From 466 to 470 Yards Long.

Width.	466 Yards Long.	467 Yards Long.	468 Yards Long.	469 Yards Long.	470 Yards Long.
Yards.	*A. R. P.*	*A. R. P.*	*A. R. P.*	*A. R. P.*	*A. R. P.*
1	0 0 15	0 0 15	0 0 15	0 0 16	0 0 16
2	0 0 31	0 0 31	0 0 31	0 0 31	0 0 31
3	0 1 6	0 1 6	0 1 6	0 1 7	0 1 7
4	0 1 22	0 1 22	0 1 22	0 1 22	0 1 22
5	0 1 37	0 1 37	0 1 37	0 1 38	0 1 38
6	0 2 12	0 2 13	0 2 13	0 2 13	0 2 13
7	0 2 28	0 2 28	0 2 28	0 2 29	0 2 29
8	0 3 3	0 3 4	0 3 4	0 3 4	0 3 4
9	0 3 19	0 3 19	0 3 19	0 3 20	0 3 20
10	0 3 34	0 3 34	0 3 35	0 3 35	0 3 35
20	1 3 28	1 3 29	1 3 29	1 3 30	1 3 31
30	2 3 22	2 3 23	2 3 24	2 3 25	2 3 26
40	3 3 16	3 3 18	3 3 19	3 3 20	3 3 21
50	4 3 10	4 3 12	4 3 14	4 3 15	4 3 17
60	5 3 4	5 3 6	5 3 8	5 3 10	5 3 12
70	6 2 38	6 3 1	6 3 3	6 3 5	6 3 8
80	7 2 32	7 2 35	7 2 38	7 3 0	7 3 3
90	8 2 26	8 2 29	8 2 32	8 2 35	8 2 38
100	9 2 20	9 2 24	9 2 27	9 2 30	9 2 34
200	19 1 1	19 1 8	19 1 14	19 1 21	19 1 27
300	28 3 21	28 3 31	29 0 1	29 0 11	29 0 21
400	38 2 2	38 2 15	38 2 28	38 3 2	38 3 15
500	48 0 22	48 0 39	48 1 16	48 1 32	48 2 9

From 471 to 475 Yards Long.

Width.	471 Yards Long.	472 Yards Long.	473 Yards Long.	474 Yards Long.	475 Yards Long.
Yards.	*A. R. P.*	*A. R. P.*	*A. R. P.*	*A. R. P.*	*A. R. P.*
1	0 0 16	0 0 16	0 0 16	0 0 16	0 0 16
2	0 0 31	0 0 31	0 0 31	0 0 31	0 0 31
3	0 1 7	0 1 7	0 1 7	0 1 7	0 1 7
4	0 1 22	0 1 22	0 1 23	0 1 23	0 1 23
5	0 1 38	0 1 38	0 1 38	0 1 38	0 1 39
6	0 2 13	0 2 14	0 2 14	0 2 14	0 2 14
7	0 2 29	0 2 29	0 2 29	0 2 30	0 2 30
8	0 3 5	0 3 5	0 3 5	0 3 5	0 3 6
9	0 3 20	0 3 20	0 3 21	0 3 21	0 3 21
10	0 3 36	0 3 36	0 3 36	0 3 37	0 3 37
20	1 3 31	1 3 32	1 3 33	1 3 33	1 3 34
30	2 3 27	2 3 28	2 3 29	2 3 30	2 3 31
40	3 3 23	3 3 24	3 3 25	3 3 27	3 3 28
50	4 3 19	4 3 20	4 3 22	4 3 23	4 3 25
60	5 3 14	5 3 16	5 3 18	5 3 20	5 3 22
70	6 3 10	6 3 12	6 3 15	6 3 17	6 3 19
80	7 3 6	7 3 8	7 3 11	7 3 14	7 3 16
90	8 3 1	8 3 4	8 3 7	8 3 10	8 3 13
100	9 2 37	9 3 0	9 3 4	9 3 7	9 3 10
200	19 1 34	19 2 1	19 2 7	19 2 14	19 2 20
300	29 0 31	29 1 1	29 1 11	29 1 21	29 1 31
400	38 3 28	39 0 1	39 0 15	39 0 28	39 1 1
500	48 2 25	48 3 2	48 3 18	48 3 35	49 0 11

From 476 to 480 Yards Long.

Width.	476 Yards Long.	477 Yards Long.	478 Yards Long.	479 Yards Long.	480 Yards Long.
Yards.	*A. R. P.*	*A. R. P.*	*A. R. P.*	*A. R. P.*	*A. R. P.*
1	0 0 16	0 0 16	0 0 16	0 0 16	0 0 16
2	0 0 31	0 0 32	0 0 32	0 0 32	0 0 32
3	0 1 7	0 1 7	0 1 7	0 1 8	0 1 8
4	0 1 23	0 1 23	0 1 23	0 1 23	0 1 23
5	0 1 39	0 1 39	0 1 39	0 1 39	0 1 39
6	0 2 14	0 2 15	0 2 15	0 2 15	0 2 15
7	0 2 30	0 2 30	0 2 31	0 2 31	0 2 31
8	0 3 6	0 3 6	0 3 6	0 3 7	0 3 7
9	0 3 22	0 3 22	0 3 22	0 3 23	0 3 23
10	0 3 37	0 3 38	0 3 38	0 3 38	0 3 39
20	1 3 35	1 3 35	1 3 36	1 3 37	1 3 37
30	2 3 32	2 3 33	2 3 34	2 3 35	2 3 36
40	3 3 29	3 3 31	3 3 32	3 3 33	3 3 35
50	4 3 27	4 3 28	4 3 30	4 3 32	4 3 33
60	5 3 24	5 3 26	5 3 28	5 3 30	5 3 32
70	6 3 21	6 3 24	6 3 26	6 3 28	6 3 31
80	7 3 19	7 3 21	7 3 24	7 3 27	7 3 29
90	8 3 16	8 3 19	8 3 22	8 3 25	8 3 28
100	9 3 14	9 3 17	9 3 20	9 3 23	9 3 27
200	19 2 27	19 2 34	19 3 0	19 3 7	19 3 14
300	29 2 1	29 2 11	29 2 20	29 2 30	29 3 0
400	39 1 14	39 1 27	39 2 1	39 2 14	39 2 27
500	49 0 28	49 1 4	49 1 21	49 1 37	49 2 14

From 481 to 485 Yards Long.

Width.	481 Yards Long.	482 Yards Long.	483 Yards Long.	484 Yards Long.	485 Yards Long.
Yards.	*A. R. P.*	*A. R. P.*	*A. R. P.*	*A. R. P.*	*A. R. P.*
1	0 0 16	0 0 16	0 0 16	0 0 16	0 0 16
2	0 0 32	0 0 32	0 0 32	0 0 32	0 0 32
3	0 1 8	0 1 8	0 1 8	0 1 8	0 1 8
4	0 1 24	0 1 24	0 1 24	0 1 24	0 1 24
5	0 2 0	0 2 0	0 2 0	0 2 0	0 2 0
6	0 2 15	0 2 16	0 2 16	0 2 16	0 2 16
7	0 2 31	0 2 32	0 2 32	0 2 32	0 2 32
8	0 3 7	0 3 7	0 3 8	0 3 8	0 3 8
9	0 3 23	0 3 23	0 3 24	0 3 24	0 3 24
10	0 3 39	0 3 39	1 0 0	1 0 0	1 0 0
20	1 3 38	1 3 39	1 3 39	2 0 0	2 0 1
30	2 3 37	2 3 38	2 3 39	3 0 0	3 0 1
40	3 3 36	3 3 37	3 3 39	4 0 0	4 0 1
50	4 3 35	4 3 37	4 3 38	5 0 0	5 0 2
60	5 3 34	5 3 36	5 3 38	6 0 0	6 0 2
70	6 3 33	6 3 35	6 3 38	7 0 0	7 0 2
80	7 3 32	7 3 35	7 3 37	8 0 0	8 0 3
90	8 3 31	8 3 34	8 3 37	9 0 0	9 0 3
100	9 3 30	9 3 33	9 3 37	10 0 0	10 0 3
200	19 3 20	19 3 27	19 3 33	20 0 0	20 0 7
300	29 3 10	29 3 20	29 3 30	30 0 0	30 0 10
400	39 3 0	39 3 14	39 3 27	40 0 0	40 0 13
500	49 2 30	49 3 7	49 3 23	50 0 0	50 0 17

From 486 to 490 Yards Long.

Width.	486 Yards Long.	487 Yards Long.	488 Yards Long.	489 Yards Long.	490 Yards Long.
Yards.	*A. R. P.*	*A. R. P.*	*A. R. P.*	*A. R. P.*	*A. R. P.*
1	0 0 16	0 0 16	0 0 16	0 0 16	0 0 16
2	0 0 32	0 0 32	0 0 32	0 0 32	0 0 32
3	0 1 8	0 1 8	0 1 8	0 1 8	0 1 9
4	0 1 24	0 1 24	0 1 25	0 1 25	0 1 25
5	0 2 0	0 2 0	0 2 1	0 2 1	0 2 1
6	0 2 16	0 2 17	0 2 17	0 2 17	0 2 17
7	0 2 32	0 2 33	0 2 33	0 2 33	0 2 33
8	0 3 9	0 3 9	0 3 9	0 3 9	0 3 10
9	0 3 25	0 3 25	0 3 25	0 3 25	0 3 26
10	1 0 1	1 0 1	1 0 1	1 0 2	1 0 2
20	2 0 1	2 0 2	2 0 3	2 0 3	2 0 4
30	3 0 2	3 0 3	3 0 4	3 0 5	3 0 6
40	4 0 3	4 0 4	4 0 5	4 0 7	4 0 8
50	5 0 3	5 0 5	5 0 7	5 0 8	5 0 10
60	6 0 4	6 0 6	6 0 8	6 0 10	6 0 12
70	7 0 5	7 0 7	7 0 9	7 0 12	7 0 14
80	8 0 5	8 0 8	8 0 11	8 0 13	8 0 16
90	9 0 6	9 0 9	9 0 12	9 0 15	9 0 18
100	10 0 7	10 0 10	10 0 13	10 0 17	10 0 20
200	20 0 13	20 0 20	20 0 26	20 0 33	20 1 0
300	30 0 20	30 0 30	30 1 0	30 1 10	30 1 20
400	40 0 26	40 1 0	40 1 13	40 1 26	40 1 39
500	50 0 33	50 1 10	50 1 26	50 2 3	50 2 19

From 491 to 495 Yards Long.

Width.	491 Yards Long.	492 Yards Long.	493 Yards Long.	494 Yards Long.	495 Yards Long.
Yards.	*A. R. P.*	*A. R. P.*	*A. R. P.*	*A. R. P.*	*A. R. P.*
1	0 0 16	0 0 16	0 0 16	0 0 16	0 0 16
2	0 0 32	0 0 33	0 0 33	0 0 33	0 0 33
3	0 1 9	0 1 9	0 1 9	0 1 9	0 1 9
4	0 1 25	0 1 25	0 1 25	0 1 25	0 1 25
5	0 2 1	0 2 1	0 2 1	0 2 2	0 2 2
6	0 2 17	0 2 18	0 2 18	0 2 18	0 2 18
7	0 2 34	0 2 34	0 2 34	0 2 34	0 2 35
8	0 3 10	0 3 10	0 3 10	0 3 11	0 3 11
9	0 3 26	0 3 26	0 3 27	0 3 27	0 3 27
10	1 0 2	1 0 3	1 0 3	1 0 3	1 0 4
20	2 0 5	2 0 5	2 0 6	2 0 7	2 0 7
30	3 0 7	3 0 8	3 0 9	3 0 10	3 0 11
40	4 0 9	4 0 11	4 0 12	4 0 13	4 0 15
50	5 0 12	5 0 13	5 0 15	5 0 17	5 0 18
60	6 0 14	6 0 16	6 0 18	6 0 20	6 0 22
70	7 0 16	7 0 19	7 0 21	7 0 23	7 0 25
80	8 0 19	8 0 21	8 0 24	8 0 26	8 0 29
90	9 0 21	9 0 24	9 0 27	9 0 30	9 0 33
100	10 0 23	10 0 26	10 0 30	10 0 33	10 0 36
200	20 1 6	20 1 13	20 1 20	20 1 26	20 1 33
300	30 1 29	30 1 39	30 2 9	30 2 19	30 2 29
400	40 2 13	40 2 26	40 2 39	40 3 12	40 3 25
500	50 2 36	50 3 12	50 3 29	51 0 5	51 0 22

From 496 to 500 Yards Long.

Width.	496 Yards Long.	497 Yards Long.	498 Yards Long.	499 Yards Long.	500 Yards Long.
Yards.	*A. R. P.*	*A. R. P.*	*A. R. P.*	*A. R. P.*	*A. R. P.*
1	0 0 16	0 0 16	0 0 16	0 0 16	0 0 17
2	0 0 33	0 0 33	0 0 33	0 0 33	0 0 33
3	0 1 9	0 1 9	0 1 9	0 1 9	0 1 10
4	0 1 26	0 1 26	0 1 26	0 1 26	0 1 26
5	0 2 2	0 2 2	0 2 2	0 2 2	0 2 3
6	0 2 18	0 2 19	0 2 19	0 2 19	0 2 19
7	0 2 35	0 2 35	0 2 35	0 2 35	0 2 36
8	0 3 11	0 3 11	0 3 12	0 3 12	0 3 12
9	0 3 28	0 3 28	0 3 28	0 3 28	0 3 29
10	1 0 4	1 0 4	1 0 5	1 0 5	1 0 5
20	2 0 8	2 0 9	2 0 9	2 0 10	2 0 11
30	3 0 12	3 0 13	3 0 14	3 0 15	3 0 16
40	4 0 16	4 0 17	4 0 19	4 0 20	4 0 21
50	5 0 20	5 0 21	5 0 23	5 0 25	5 0 26
60	6 0 24	6 0 26	6 0 28	6 0 30	6 0 32
70	7 0 28	7 0 30	7 0 32	7 0 35	7 0 37
80	8 0 32	8 0 34	8 0 37	8 1 0	8 1 2
90	9 0 36	9 0 39	9 1 2	9 1 5	9 1 8
100	10 1 0	10 1 3	10 1 6	10 1 10	10 1 13
200	20 1 39	20 2 6	20 2 13	20 2 19	20 2 26
300	30 2 39	30 3 9	30 3 19	30 3 29	30 3 39
400	40 3 39	41 0 12	41 0 25	41 0 38	41 1 12
500	51 0 38	51 1 15	51 1 31	51 2 8	51 2 24

A TABLE

SHOWING

THE WIDTH REQUIRED

FOR AN

ACRE OF LAND,

FROM

ONE TO FIVE HUNDRED YARDS IN LENGTH.

EXPLANATION OF THE TABLE.

First, find the length of the piece in yards in the table, and opposite it stands the width, in yards, feet, and inches, required for an acre corresponding with that length:—Thus, if the length be 278 yards, then against it is 17 yards, 1 foot, 3 inches—the width to make an acre.

The width for two, three, or more acres, may be found by doubling, trebling, &c., the width for one acre; also, the width for half or a quarter of an acre is found by taking the half or the quarter of the width for an acre.

Suppose the length of a square piece of land is 162 yards, then the width for an acre is 29 yards, 2 feet, 8 inches, the double of which is 59 yards, 2 feet, 4 inches, the width for two acres. The treble of it is 89 yards, 2 feet, 0 inches, the width for three acres.—Also, the half of it is 14 yards, 2 feet, 10 inches, the width for half an acre. The quarter of it is 7 yards, 1 foot, 5 inches, the width for a quarter of an acre.

The width for three quarters of an acre is found by taking the width for half and a quarter of an acre, and adding them together.

From 1 to 100 Yards in Length.

Leng.	Width.			Leng.	Width.			Leng.	Width.			Leng.	Width.		
Yds.	*Yds.*	*Ft.*	*In.*	*Yds.*	*Yds.*	*Ft.*	*In.*	*Yds.*	*Yds.*	*Ft.*	*In.*	*Yds.*	*Yds.*	*Ft.*	*In.*
1	4840	0	0	26	186	0	6	51	94	2	9	76	63	2	1
2	2420	0	0	27	179	0	10	52	93	0	3	77	62	2	7
3	1613	1	0	28	172	2	7	53	91	1	0	78	62	0	2
4	1210	0	0	29	166	2	9	54	89	1	11	79	61	0	10
5	968	0	0	30	161	1	0	55	88	0	0	80	60	1	6
6	806	2	0	31	156	0	5	56	86	1	4	81	59	2	4
7	691	1	4	32	151	0	9	57	84	2	9	82	59	0	1
8	605	0	0	33	146	2	0	58	83	1	5	83	58	1	0
9	537	2	4	34	142	1	1	59	82	0	2	84	57	1	11
10	484	0	0	35	138	0	11	60	80	2	0	85	56	2	10
11	440	0	0	36	134	1	4	61	79	1	1	86	56	0	11
12	403	1	0	37	130	2	6	62	78	0	2	87	55	1	11
13	372	1	0	38	127	1	2	63	76	2	6	88	55	0	0
14	345	2	2	39	124	0	4	64	75	1	11	89	54	1	2
15	322	2	0	40	121	0	0	65	74	1	5	90	53	2	4
16	302	1	6	41	118	0	2	66	73	1	0	91	53	0	7
17	284	2	2	42	115	0	9	67	72	0	9	92	52	1	10
18	268	2	8	43	112	1	9	68	71	0	7	93	52	0	2
19	254	2	3	44	110	0	0	69	70	0	6	94	51	1	6
20	242	0	0	45	107	1	8	70	69	0	5	95	50	2	11
21	230	1	6	46	105	0	8	71	68	0	7	96	50	1	3
22	220	0	0	47	103	0	0	72	67	0	8	97	49	2	9
23	210	1	4	48	100	2	6	73	66	0	11	98	49	1	2
24	201	2	0	49	98	2	4	74	65	1	3	99	48	2	8
25	193	1	10	50	96	2	5	75	64	1	8	100	48	1	3

From 101 to 200 Yards in Length.

Leng.	Width.			Leng.	Width.			Leng.	Width.			Leng.	Width.		
Yds.	*Yds.*	*Ft.*	*In.*	*Yds.*	*Yds.*	*Ft.*	*In.*	*Yds.*	*Yds.*	*Ft.*	*In.*	*Yds.*	*Yds.*	*Ft.*	*In.*
101	47	2	10	126	38	1	3	151	32	0	2	176	27	1	6
102	47	1	5	127	38	0	4	152	31	2	7	177	27	1	1
103	47	0	0	128	37	2	6	153	31	1	11	178	27	0	7
104	46	1	8	129	37	1	7	154	31	1	4	179	27	0	2
105	46	0	4	130	37	0	9	155	31	0	9	180	26	2	8
106	45	2	0	131	36	2	11	156	31	0	1	181	26	2	3
107	45	0	9	132	36	2	0	157	30	2	6	182	26	1	10
108	44	2	6	133	36	1	3	158	30	1	11	183	26	1	5
109	44	1	3	134	36	0	5	159	30	1	4	184	26	0	11
110	44	0	0	135	35	2	7	160	30	0	9	185	26	0	6
111	43	1	10	136	35	1	10	161	30	0	3	186	26	0	1
112	43	0	8	137	35	1	0	162	29	2	8	187	25	2	8
113	42	2	6	138	35	0	3	163	29	2	1	188	25	2	3
114	42	1	5	139	34	2	6	164	29	1	7	189	25	1	10
115	42	0	4	140	34	1	9	165	29	1	0	190	25	1	6
116	41	2	3	141	34	1	0	166	29	0	6	191	25	1	1
117	41	1	2	142	34	0	4	167	29	0	0	192	25	0	8
118	41	0	1	143	33	2	7	168	28	2	6	193	25	0	3
119	40	2	1	144	33	1	10	169	28	2	0	194	24	2	11
120	40	1	0	145	33	1	2	170	28	1	5	195	24	2	6
121	40	0	0	146	33	0	6	171	28	0	11	196	24	2	1
122	39	2	1	147	32	2	10	172	28	0	6	197	24	1	9
123	39	1	1	148	32	2	2	173	28	0	0	198	24	1	4
124	39	0	1	149	32	1	6	174	27	2	6	199	24	1	0
125	38	2	2	150	32	0	10	175	27	2	0	200	24	0	8

From 201 to 300 Yards in Length.

Leng.	Width.			Leng.	Width.			Leng.	Width.			Leng.	Width.		
Yds.	Yds.	Ft.	In.	Yds.	Yds.	Ft.	In.	Yds.	Yds.	Ft.	In.	Yds.	Yds.	Ft.	In.
201	24	0	3	226	21	1	3	251	19	0	11	276	17	1	8
202	23	2	11	227	21	1	0	252	19	0	8	277	17	1	5
203	23	2	7	228	21	0	9	253	19	0	5	278	17	1	3
204	23	2	3	229	21	0	5	254	19	0	2	279	17	1	1
205	23	1	10	230	21	0	2	255	19	0	0	280	17	0	11
206	23	1	6	231	20	2	11	256	18	2	9	281	17	0	9
207	23	1	2	232	20	2	8	257	18	2	6	282	17	0	6
208	23	0	10	233	20	2	4	258	18	2	4	283	17	0	4
209	23	0	6	234	20	2	1	259	18	2	1	284	17	0	2
210	23	0	2	235	20	1	10	260	18	1	11	285	17	0	0
211	22	2	10	236	20	1	7	261	18	1	8	286	16	2	10
212	22	2	6	237	20	1	4	262	18	1	6	287	16	2	8
213	22	2	3	238	20	1	1	263	18	1	3	288	16	2	5
214	22	1	11	239	20	0	10	264	18	1	0	289	16	2	3
215	22	1	7	240	20	0	6	265	18	0	10	290	16	2	1
216	22	1	3	241	20	0	3	266	18	0	8	291	16	1	11
217	22	0	11	242	20	0	0	267	18	0	5	292	16	1	9
218	22	0	8	243	19	2	10	268	18	0	3	293	16	1	7
219	22	0	4	244	19	2	7	269	18	0	0	294	16	1	5
220	22	0	0	245	19	2	4	270	17	2	10	295	16	1	3
221	21	2	9	246	19	2	1	271	17	2	7	296	16	1	1
222	21	2	5	247	19	1	10	272	17	2	5	297	16	0	11
223	21	2	2	248	19	1	7	273	17	2	3	298	16	0	9
224	21	1	10	249	19	1	4	274	17	2	0	299	16	0	7
225	21	1	7	250	19	1	1	275	17	1	10	300	16	0	5

From 301 to 400 Yards in Length.

Leng.	Width.			Leng.	Width.			Leng.	Width.			Leng.	Width.		
Yds.	*Yds.*	*Ft.*	*In.*	*Yds.*	*Yds.*	*Ft.*	*In.*	*Yds.*	*Yds.*	*Ft.*	*In.*	*Yds.*	*Yds.*	*Ft.*	*In.*
301	16	0	3	326	14	2	7	351	13	2	5	376	12	2	8
302	16	0	1	327	14	2	5	352	13	2	8	377	12	2	7
303	16	0	0	328	14	2	4	353	13	2	2	378	12	2	5
304	15	2	10	329	14	2	2	354	13	2	1	379	12	2	4
305	15	2	8	330	14	2	0	355	13	1	11	380	12	2	3
306	15	2	6	331	14	1	11	356	13	1	10	381	12	2	2
307	15	2	4	332	14	1	9	357	13	1	9	382	12	2	1
308	15	2	2	333	14	1	8	358	13	1	7	383	12	1	11
309	15	2	0	334	14	1	6	359	13	1	6	384	12	1	10
310	15	1	11	335	14	1	5	360	13	1	4	385	12	1	9
311	15	1	9	336	14	1	3	361	13	1	3	386	12	1	8
312	15	1	7	337	14	1	2	362	13	1	2	387	12	1	7
313	15	1	5	338	14	1	0	363	13	1	0	388	12	1	6
314	15	1	3	339	14	0	10	364	13	0	11	389	12	1	4
315	15	1	2	340	14	0	9	365	13	0	10	390	12	1	3
316	15	1	0	341	14	0	7	366	13	0	9	391	12	1	2
317	15	0	10	342	14	0	6	367	13	0	7	392	12	1	1
318	15	0	8	343	14	0	4	368	13	0	6	393	12	1	0
319	15	0	7	344	14	0	3	369	13	0	5	394	12	0	11
320	15	0	5	345	14	0	2	370	13	0	3	395	12	0	10
321	15	0	3	346	14	0	0	371	13	0	2	396	12	0	8
322	15	0	2	347	13	2	11	372	13	0	1	397	12	0	7
323	15	0	0	348	13	2	9	373	13	0	0	398	12	0	6
324	14	2	10	349	13	2	8	374	12	2	10	399	12	0	5
325	14	2	9	350	13	2	6	375	12	2	9	400	12	0	4

From 401 to 500 Yards in Length.

Leng.	Width.			Leng.	Width.			Leng.	Width.			Leng.	Width.		
Yds.	Yds.	Ft.	In.	Yds.	Yds.	Ft.	In.	Yds.	Yds.	Ft.	In.	Yds.	Yds.	Ft.	In.
401	12	0	3	426	11	1	2	451	10	2	3	476	10	0	7
402	12	0	2	427	11	1	1	452	10	2	2	477	10	0	6
403	12	0	1	428	11	1	0	453	10	2	1	478	10	0	5
404	12	0	0	429	11	0	11	454	10	2	0	479	10	0	4
405	11	2	11	430	11	0	10	455	10	1	11	480	10	0	3
406	11	2	10	431	11	0	9	456	10	1	11	481	10	0	3
407	11	2	9	432	11	0	8	457	10	1	10	482	10	0	2
408	11	2	8	433	11	0	7	458	10	1	9	483	10	0	1
409	11	2	7	434	11	0	6	459	10	1	8	484	10	0	0
410	11	2	5	435	11	0	5	460	10	1	7	485	10	0	0
411	11	2	4	436	11	0	4	461	10	1	6	486	9	2	11
412	11	2	3	437	11	0	3	462	10	1	6	487	9	2	10
413	11	2	2	438	11	0	2	463	10	1	5	488	9	2	10
414	11	2	1	439	11	0	1	464	10	1	4	489	9	2	9
415	11	2	0	440	11	0	0	465	10	1	3	490	9	2	8
416	11	1	11	441	11	0	0	466	10	1	2	491	9	2	7
417	11	1	10	442	10	2	11	467	10	1	2	492	9	2	7
418	11	1	9	443	10	2	10	468	10	1	1	493	9	2	6
419	11	1	8	444	10	2	9	469	10	1	0	494	9	2	5
420	11	1	7	445	10	2	8	470	10	0	11	495	9	2	4
421	11	1	6	446	10	2	7	471	10	0	10	496	9	2	4
422	11	1	5	447	10	2	6	472	10	0	10	497	9	2	3
423	11	1	4	448	10	2	5	473	10	0	9	498	9	2	2
424	11	1	3	449	10	2	5	474	10	0	8	499	9	2	2
425	11	1	2	450	10	2	4	475	10	0	7	500	9	2	1

TABLES

FOR

MANURING LAND.

EXPLANATION OF THE FIRST TABLE.

The left-hand column shows the distance of the heaps of manure in yards, the figures at the top the number of heaps in a load, and under them the number of *loads* required for an acre for any given distance of the heaps: —thus, if heaps of clay are set 3½ yards asunder, and 5 heaps made of a load, then under five in the table, and opposite 3½ yards in the width, you will find 80, which shows the number of loads required for an acre. Again, if the heaps of dung are 7 yards distant from each other, and 8 heaps made of a load, the table shows that loads are required to manure an acre.

First Table for Manuring Land.

Distance of the Heaps.	1 Heap in a Load.	2 Heaps in a Load.	3 Heaps in a Load.	4 Heaps in a Load.	5 Heaps in a Load.	6 Heaps in a Load.	7 Heaps in a Load.	8 Heaps in a Load.	9 Heaps in a Load.	10 Heaps in a Load.
Yards.	Loads.	Loads.	Loads.	Loads.	Loads.	Loads.	Loads.	Loads.	Loads.	Loads.
1	4840	2420	1614	1210	968	807	692	605	538	484
1½	2152	1076	718	538	431	359	308	269	240	216
2	1210	605	404	303	242	202	173	152	135	121
2½	775	388	259	194	155	130	111	97	87	78
3	538	269	180	135	108	90	77	68	60	54
3½	396	198	132	99	80	66	57	50	44	40
4	303	152	101	76	61	51	44	38	34	31
4½	240	120	80	60	48	40	35	30	27	24
5	294	97	65	49	39	33	28	25	22	20
5½	160	80	54	40	32	27	23	20	18	16
6	135	68	45	34	27	23	20	17	15	14
6½	115	58	39	29	23	20	17	15	13	12
7	99	50	33	25	20	17	15	13	11	10
7½	87	44	29	22	18	15	13	11	10	9
8	76	38	26	19	16	13	11	10	9	8
8½	67	34	23	17	14	12	10	9	8	7
9	60	30	20	15	12	10	9	8	7	6
9½	54	27	18	14	11	9	8	7	6	6
10	49	25	17	13	10	9	7	7	6	5

EXPLANATION

OF THE

SECOND TABLE FOR MANURING LAND.

The first of these tables is made on the supposition that the distance of the rows of manure is equal to the distance of the heaps from each other; but as this is not always the case, particularly in arable lands, where the furrows are visible, the following table will show the number of *heaps* required to cover an acre, for any given width of the rows and distance of the heaps from each other.

Having found the width of the rows in the figures placed at the top of the columns, and the distance of the heaps in the column on the left-hand side; then immediately under the former, and opposite the latter, stands the number of heaps required for an acre:—Thus, if the rows are 7½ yards wide, and the heaps 6 yards from each other, then under 7½ in the table, and opposite 6, we find 108, the number of heaps for an acre; and if this number be divided by the number of heaps made of a load, it will then show the number of loads laid on an acre.

In like manner, the number of heaps on an acre is found by the table for any other given width of the rows and distance of the heaps.

In the foregoing table a *less* number of loads, and in the following table a *less* number of heaps will not cover an acre; although, in some instances, the numbers set down in the tables will cover a small quantity more than an acre

Second Table for Manuring Land.

Distance of the Heaps.	Rows 1½ Yards Wide.	Rows 2 Yards Wide.	Rows 2½ Yards Wide.	Rows 3 Yards Wide.	Rows 3½ Yards Wide.	Rows 4 Yards Wide.	Rows 4½ Yards Wide.	Rows 5 Yards Wide.	Rows 5½ Yards Wide.
Yards.	Heaps.	Heaps.	Heaps.	Heaps.	Heaps.	Heaps.	Heaps.	Heaps.	Heaps.
1½	2152	1614	1291	1076	922	807	718	646	587
2	1614	1210	968	807	692	605	538	484	440
2½	1229	968	775	646	554	484	431	388	335
3	1076	807	646	538	461	404	359	323	294
3½	922	692	554	461	396	346	308	277	252
4	807	605	484	404	346	303	269	242	220
4½	718	538	431	359	308	269	240	216	196
5	646	484	388	323	277	242	216	194	176
5½	587	440	352	294	252	220	196	176	160
6	538	404	323	269	231	202	180	162	147
6½	497	373	298	249	213	187	166	149	136
7	461	346	277	231	198	173	154	139	126
7½	431	323	259	216	185	162	144	130	118
8	404	303	242	202	173	152	135	121	110
8½	380	285	228	190	163	143	127	114	104
9	359	269	216	180	154	135	120	108	98
9½	340	255	204	170	146	128	114	102	93
10	323	242	194	162	139	121	108	97	88

Second Table for Manuring Land.

Distance of the Heaps.	Rows 6 Yards Wide.	Rows 6½ Yards Wide.	Rows 7 Yards Wide.	Rows 7½ Yards Wide.	Rows 8 Yards Wide.	Rows 8½ Yards Wide.	Rows 9 Yards Wide.	Rows 9½ Yards Wide.	Rows 10 Yards Wide.
Yards.	Heaps.	Heaps.	Heaps.	Heaps.	Heaps.	Heaps.	Heaps.	Heaps.	Heaps.
1½	538	497	461	431	404	380	359	340	323
2	404	373	346	323	303	285	269	255	242
2½	323	298	277	259	242	228	216	204	194
3	269	249	231	216	202	190	180	170	162
3½	231	213	198	185	173	163	154	146	139
4	202	187	173	162	152	143	135	128	121
4½	180	166	154	144	135	127	120	114	108
5	162	149	139	130	121	114	108	102	97
5½	147	136	126	118	110	104	98	93	88
6	135	125	116	108	101	95	90	85	81
6½	125	115	107	100	94	88	83	79	75
7	116	107	99	93	87	82	77	73	70
7½	108	100	93	87	81	76	72	68	65
8	101	94	87	81	76	72	68	64	61
8½	95	88	82	76	72	67	64	60	57
9	90	83	77	72	68	64	60	57	54
9½	85	79	73	68	64	60	57	54	51
10	81	75	70	65	61	57	54	51	49

TABLE showing the number of Plants required per Acre at given distances.

Feet.		Ft. In.	Plants.	Ft. In.		Ft. In.	Plants.	Ft. In.		Ft. In.	Plants.	Ft. In.		Ft. In.	Plants.
9	by	9	537	5 6	by	3 9	2112	4 6	by	2 0	4840	3 3	by	3 3	4124
do.	"	8	605	do.	"	3 6	2262	do.	"	1 9	5531	do.	"	3 0	4818
do.	"	7	691	do.	"	3 3	2436	do.	"	1 6	6453	do.	"	2 9	4873
do.	"	6	806	do.	"	3 0	2640	do.	"	1 3	7744	do.	"	2 6	5361
do.	"	5	968	do.	"	2 9	2880	do.	"	1 0	9680	do.	"	2 3	5956
8	"	8	680	do.	"	2 6	3168	4 3	"	4 3	2411	do.	"	2 0	6701
do.	"	7	777	do.	"	2 3	3520	do.	"	4 0	2562	do.	"	1 9	7658
do.	"	6	905	do.	"	2 0	3960	do.	"	3 9	2733	do.	"	1 6	8935
do.	"	5	1089	do.	"	1 9	4525	do.	"	3 6	2914	do.	"	1 3	10722
do.	"	4	1361	do.	"	1 6	5280	do.	"	3 3	3153	do.	"	1 0	13403
do.	"	3	1815	do.	"	1 3	6336	do.	"	3 0	3416	3	"	3 0	4840
7	"	7 0	888	do.	"	1 0	7920	do.	"	2 9	3727	do.	"	2 9	5289
do.	"	6 6	957	5	"	5 0	1742	do.	"	2 6	4099	do.	"	2 6	5808
do.	"	6 0	1037	do.	"	4 9	1834	do.	"	2 3	4555	do.	"	2 3	6453
do.	"	5 6	1131	do.	"	4 6	1936	do.	"	2 0	5124	do.	"	2 0	7260
do.	"	5 0	1244	do.	"	4 3	2049	do.	"	1 9	5856	do.	"	1 9	8297
do.	"	4 6	1382	do.	"	4 0	2178	do.	"	1 6	6832	do.	"	1 6	9680
do.	"	4 0	1555	do.	"	3 9	2323	do.	"	1 3	8199	do.	"	1 3	11616
do.	"	3 6	1777	do.	"	3 6	2489	do.	"	1 0	10249	do.	"	1 0	14520
do.	"	3 0	2074	do.	"	3 3	2680	4	"	4 0	2722	2 9	"	2 9	5760
do.	"	2 6	2489	do.	"	3 0	2904	do.	"	3 9	2904	do.	"	2 6	6336
do.	"	2 0	3111	do.	"	2 9	3168	do.	"	3 6	3111	do.	"	2 3	7040
do.	"	1 6	4148	do.	"	2 6	3484	do.	"	3 3	3350	do.	"	2 0	7920
do.	"	1 0	6222	do.	"	2 3	3872	do.	"	3 0	3630	do.	"	1 9	9051
6 6	"	6 6	1031	do.	"	2 0	4356	do.	"	2 9	3960	do.	"	1 6	10560
do.	"	6 0	1116	do.	"	1 9	4978	do.	"	2 6	4356	do.	"	1 3	12670
do.	"	5 6	1218	do.	"	1 6	5808	do.	"	2 3	4840	do.	"	1 0	15840
do.	"	5 0	1340	do.	"	1 3	6969	do.	"	2 0	5445	2 6	"	2 6	6969
do.	"	4 6	1489	do.	"	1 0	8712	do.	"	1 9	6222	do.	"	2 3	7740
do.	"	4 0	1675	4 9	"	4 9	1930	do.	"	1 6	7260	do.	"	2 0	8712
do.	"	3 6	1914	do.	"	4 6	2037	do.	"	1 3	8712	do.	"	1 9	9956
do.	"	3 0	2233	do.	"	4 3	2157	do.	"	1 0	10890	do.	"	1 6	11616
do.	"	2 6	2680	do.	"	4 0	2292	3 9	"	3 9	3097	do.	"	1 3	13939
do.	"	2 0	3350	do.	"	3 9	2445	do.	"	3 6	3318	do.	"	1 0	17424
do.	"	1 6	4467	do.	"	3 6	2620	do.	"	3 3	3574	2 3	"	2 3	8604
do.	"	1 0	6701	do.	"	3 3	2821	do.	"	3 0	3872	do.	"	2 0	9680
6	"	6 0	1210	do.	"	3 0	3056	do.	"	2 9	4224	do.	"	1 9	11062
do.	"	5 9	1262	do.	"	2 9	3334	do.	"	2 6	4646	do.	"	1 6	12906
do.	"	5 6	1320	do.	"	2 6	3668	do.	"	2 3	5162	do.	"	1 3	15488
do.	"	5 0	1452	do.	"	2 3	4075	do.	"	2 0	5808	do.	"	1 0	19360
do.	"	4 6	1613	do.	"	2 0	4585	do.	"	1 9	6637	2	"	2 0	10890
do.	"	4 0	1815	do.	"	1 9	5248	do.	"	1 6	7744	do.	"	1 9	12445
do.	"	3 6	2074	do.	"	1 6	6113	do.	"	1 3	9272	do.	"	1 6	14520
do.	"	3 0	2420	do.	"	1 3	7336	do.	"	1 0	11616	do.	"	1 3	17424
do.	"	2 6	2904	do.	"	1 0	9170	3 6	"	3 6	3555	do.	"	1 0	21780
do.	"	2 0	3630	4 6	"	4 6	2151	do.	"	3 3	3829	1 9	'	1 9	14223
do.	"	1 6	4840	do.	"	4 3	2277	do.	"	3 0	4148	do.	"	1 6	16594
do.	"	1 0	7260	do.	"	4 0	2420	do.	"	2 9	4525	do.	"	1 3	19913
5 6	"	5 6	1417	do.	"	3 9	2581	do.	"	2 6	4978	do.	"	1 0	24454
do.	"	5 3	1508	do.	"	3 6	2765	do.	"	2 3	5531	1 6	"	1 6	19360
do.	"	5 0	1584	do.	"	3 3	2978	do.	"	2 0	6222	do.	"	1 3	23232
do.	"	4 9	1667	do.	"	3 0	3226	do.	"	1 9	7111	do.	"	1 0	29040
do.	"	4 6	1760	do.	"	2 9	3520	do.	"	1 6	8297	1 3	"	1 3	27878
do.	"	4 3	1863	do.	"	2 6	3872	do.	"	1 3	9956	do.	"	1 0	34848
do.	"	4 0	1980	do.	"	2 3	4302	do.	"	1 0	12445	1	"	1 0	43560

PLOUGHING.

Names of Fields.	Length of land.	Breadth to give an acre.	Breadth of the furrow slice.	Number of furrows in an acre.	Time that it takes in turning.	Time taken in turning the soil.	Number of hours in the day's work.
	Yards.	Yards.	Inches.		H. M.	H. M.	Hours.
Short lands...	78	186	8	279	4 39	3 21	8
Harper's Hill.	149	98	8	147	2 27	5 33	8
Home Close ..	200	73	8	109	1 49	6 11	8
East Lake ...	212	69	8	103	1 43	6 17	8
Long lands. .	274	53	8	79	1 19	6 41	8

EXPLANATION.

When the land is no more than 78 yards long, 4 hours and 39 minutes are spent merely in turning at the ends, in a journey of 8 hours; whereas, when the land is 274 yards long, 1 hour and 19 minutes are sufficient for that purpose in the same length of time.

The whole series of furrows on an acre of land, supposing each to be 9 inches in width, would extend in length to 19,360 yards; and, adding 12 yards to every 220, for the average estimated ground travelled over in turning, the whole work of ploughing one acre may be given as extending to 20,416 yards, or 11 miles and nearly five furlongs.

OVERSEER'S ACCOUNT OF TIME.

The time of any number of workmen may be kept with most perfect accuracy, and in the simplest manner, by the following mode:

Procure a slate, and mark upon it by means of a sharp-pointed instrument, so that they shall be deeply indented, the following straight lines and crosses. The workmen's names, together with the rate of wages, are to be added with pencil; and on the evening of each day, the filling up of segments of circles with the same must take place, according to the time each man has been employed, together with any remarks that may follow. At the end of the week every man's time is thus ascertained, without the possibility of mistake or misunderstanding; and after all is settled, the application of a wet sponge makes ready for another week, the crosses remaining indented in the slate.

Workmen.	M.	T.	W.	T.	F.	S.	Per day.	No. of days.	Amount	Remarks.
J. Smith..							$0 75	$2\frac{1}{4}$	$1 68$\frac{3}{4}$	Attentive and careful.
T. Webb..							1 00	$2\frac{1}{2}$	2 50	Lent him $2 25
J. Walker							1 00	4	4 00	
J. Frith...							1 25	$2\frac{1}{2}$	3 12$\frac{1}{2}$	Lent him $2 75
G. Evans.							0 75	3	2 25	
F. Small..							1 12$\frac{1}{2}$	$3\frac{1}{2}$	3 93$\frac{3}{4}$	Work ill done.
W. Green.								·		Absent, sick.

Whole day employed.
Forenoon of the day.
Afternoon of the day.
First quarter.
Second quarter.
Third quarter.
Fourth quarter.
First and third quarters.
Second and fourth quarters.
First and fourth quarters.
Second and third quarters.
First three quarters.
Last three quarters.
First, third and fourth quarters
First, second and fourth quarters.
Whole day idle.

THE RIGHT-ANGLE.

Every Farmer ought to know before he erects a house or any other building, how to lay off the sills in an exact square. He must measure off 8 feet from the end of one sill, and there make a mark; he then measures off 6 feet on the sill lying at right angles with the first, and makes another mark; he then lays on a 10 feet pole, one end of it squaring with the first mark—and if the other end of it does not exactly meet the second mark, he moves the sill in or out, until it exactly squares with it. The figure which he thus makes in marking off his sills and in laying down his 10 feet pole, is *a right-angled triangle*, A, B, C; the right-angle or square corner being at B.

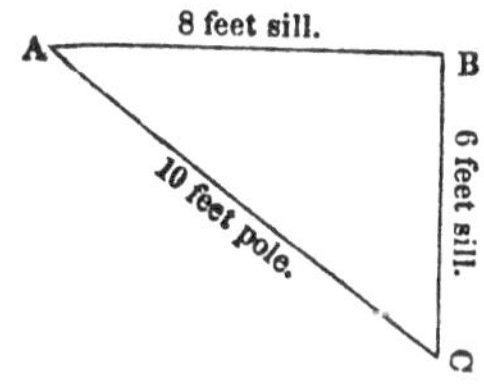

Now, unless the line A C be 10 feet long when the other two are 8 and 6 respectively, the corner B will not be a *square corner;* for it is found by mathematicians, that in every right-angled triangle, the longest line—the line opposite to the right-angle—when squared, is just equal to the squares of the other two, the lines A B and B C. This is the reason why carpenters adopt this rule to lay their sills square—8 squared being 64, and 6 squared, 36; both together being 100—ten squared being 100, ten times ten.

MEASUREMENT OF CORN IN THE CRIB.

After levelling the corn, multiply the length and breadth of the house together, and the product by the depth, which will give the cubic feet of the bulk of corn: then divide this last product by 12, and the quotient will be the number of barrels of shelled corn contained in the house or crib. If there be a remainder after the division, it will be so many twelfths of a barrel of shelled corn over.

Example.

```
    12 feet long
    11 feet broad
   ----
   132
     6 feet deep
   ----
12)792 cubic feet
   ----
    66 barrels of shelled corn
     5 bushels in a barrel
   ----
   330 bushels of shelled corn.
```

Memoranda.—21,500 cubic inches will contain ten bushels of shelled corn, but the same space filled with corn *in the ear* will shell out rather more than five bushels. These 21,500 cubic inches contain 12 cubic feet, and 764 cubic inches over. Now, two barrels, or ten bushels in the ear, will generally, in shelling, overrun just about these 764 cubic inches.

THE SPAN LEVEL, FOR DRAINING.

To fit the instrument for giving descent to a drain, place it so that the plummet or bob-line shall fall into the mark on the brace-piece; then place a block an inch thick under one of the feet, and mark the place where the line crosses the brace; then add another inch block, and mark again, repeating the operation as often as it is wished; then take out the blocks, and go over the process again, by placing them one at a time under the contrary foot, making the necessary marks on the brace; the lines should then be numbered 1, 2, 3, &c. from the middle. Now, when the Level is so placed that the bob-line falls on mark No. 1, or that nearest the *middle line*, it is evident that the end of the instrument nearest the bob-line will be one inch lower than the other; and when on No. 2, two inches, &c. And if the span of the feet be made 16 feet 6 inches, or one perch, and the lines marked as above, the plummet-line, resting on mark No. 1, shows a fall or rise of one inch to the perch, which is 26 feet 8 inches to the mile; when resting on mark No. 2, two inches to the perch, &c. And if the span be 12 feet 4½ inches, or three-quarters of a perch, the blocks used for raising in marking must be three-quarters of an inch thick, then each of the lines will show the same fall as above. The Level as above divided, will show the fall in certain proportions to a given length; but there are cases in draining where these relative proportions do not exist on the ground: thus, the above lines and their subdivisions would show a rise or fall of 6 ft. 8 in. 13 ft. 4 in. or 26 ft. 8 in. to the mile, and sundry others, by various additions of them together; but if the fall of the ground were found to be any number of feet and inches between these, it could not be properly proportioned in the drains by the use of the above lines; but a ready way to obviate this difficulty, is to make the span 12 feet 6 inches, the 8th part of an 100 feet; then ascertain the number of inches of fall in 100 feet of the ground to be drained, and for every inch let one-eighth of an inch be allowed to the block used in raising the feet when making the slope marks; or let the block be as many eighths of an inch thick as there are inches fall in 100 feet—thus, if the fall in 100 feet be 6 inches, the block must be 6-8 or ¾ of an inch; if 7 inches, ⅞; 8 inches, 1 inch, &c. An instrument of this length is more convenient than any other, and can be, by the above method, at any time adapted to any case that might occur; the principle being simply to find the fall of a portion of the ground to be drained equal in length to the span of the instrument, and apply a block in making the marks equal in thickness to that fall.

A TABLE

SHOWING HOW MUCH THE

BUSHEL IN COMMON USE CONTAINS

MORE OR LESS (IN PINTS),

THAN THE

STANDARD IMPERIAL BUSHEL.

THE IMPERIAL BUSHEL.

The various disputes which frequently arise respecting the measure of grain, by reason of the great diversity of form of the bushel in common use amongst merchants, farmers, millers, and others, rendering it oftentimes next to impossible to know correctly whether it be too large or too small—one being quite as likely to happen as the other, even in the hands of well-disposed persons—the following table has been drawn up for the purpose of settling the question. By reference to this, any one can satisfy himself of the capacity of his own or his neighbour's measure, while in the act of buying or selling, with the greatest exactitude. The form of the bushel was originally 19½ inches in diameter, and of sufficient internal capacity to hold eight gallons; but it was found extremely inconvenient for the purpose of measuring grain into bags, &c. being too broad and shallow; the bushels in common use are narrower, and in proportion deeper, so that the exact content shall be the same, be the depth and diameter what they may. But as no one could easily conceive the great and serious difference that even the eighth part of an inch in diameter in the bushel he uses will cause him, it is of importance that some ready mode should be adopted, by which every man may satisfy himself on the subject; and it is with this view the following table is given, which shows the content of any bushel in pints and quarters of pints, of all the various diameters and depths that bushels in common use are to be met with: and as a bushel ought to contain exactly 64 pints, whatever the content in the table exceeds or falls short of that sum, will be what the bushel holds more or less than its proper quantity.

Explanation. The first column contains the diameter of the bushel in inches and parts of an inch; the second the depth, and the third its contents in pints. Thus, if a bushel be 15 inches in diameter, and 12½ inches in depth, its content will be 63¾ pints, or one-quarter of a pint less than it ought to be. Again, if a bushel be 15½ inches in diameter, and 12 inches in depth, its content will be 65¼ pints, or a pint and a quarter more than measure.

THE IMPERIAL BUSHEL TABLE.

Diameter in Inches.	Depth in Inches.	Content in Pints.	Diameter in Inches.	Depth in Inches.	Content in Pints.	Diameter in Inches.	Depth in Inches.	Content in Pints.	Diameter in Inches.	Depth in Inches.	Content in Pints.
13	16⅜	62¾	13½	15⅛	62½	14	14⅛	62¾	14½	13⅛	62½
	16½	63¼		15¼	63		14¼	63¼		13¼	63
	16⅝	63¾		15⅜	63½		14⅜	63¾		13⅜	63¾
	16¾	64¼		15½	64		14½	64½		13½	64¼
	16⅞	64½		15⅝	64½		14⅝	65		13⅝	65
	17	65		15¾	65		14¾	65½		13¾	65½
13⅛	16	62½	13⅝	14⅞	62½	14⅛	13⅞	62¾	14⅝	12⅞	62½
	16⅛	63		15	63		14	63¼		13	63
	16¼	63½		15⅛	63¾		14⅛	63¾		13⅛	63½
	16⅜	64		15¼	64¼		14¼	64½		13¼	64¼
	16½	64½		15⅜	64¾		14⅜	65		13⅜	64¾
	16⅝	65		15½	65¼		14½	65½		13½	65½
13¼	15¾	62¾	13¾	14⅝	62¾	14¼	13⅝	62¾	14¾	12⅝	62¼
	15⅞	63¼		14¾	63¼		13¾	63¼		12¾	62¾
	16	63¾		14⅞	63¾		13⅞	63¾		12⅞	63½
	16⅛	64¼		15	64¼		14	64½		13	64
	16¼	64¾		15⅛	64¾		14⅛	65		13⅛	64¾
	16⅜	65¼		15¼	65¼		14¼	65½		13¼	65¼
13⅜	15½	62¾	13⅞	14⅜	62¾	14⅜	13⅜	62½	14⅞	12⅜	62
	15⅝	63¼		14½	63¼		13½	63¼		12½	62¾
	15¾	63¾		14⅝	63¾		13⅝	63¾		12⅝	63¼
	15⅞	64¼		14¾	64¼		13¾	64½		12¾	64
	16	64¾		14⅞	65		13⅞	65		12⅞	64½
	16⅛	65¼		15	65½		14	65½		13	65¼

THE IMPERIAL BUSHEL TABLE.

Diameter in Inches.	Depth in Inches.	Content in Pints.	Diameter in Inches.	Depth in Inches.	Content in Pints.	Diameter in Inches.	Depth in Inches.	Content in Pints.	Diameter in Inches.	Depth in Inches.	Content in Pints.
15	12¼	62½	15½	11⅜	62	16	10¾	62¼	16½	10	61¾
	12⅜	63		11½	62½		10⅞	63		10⅛	62½
	12½	63¾		11⅝	63¼		11	63¾		10¼	63¼
	12⅝	64¼		11¾	64		11⅛	64½		10⅜	64
	12¾	65		11⅞	64¾		11¼	65¼		10½	64¾
	12⅞	65¾		12	65¼		11⅜	66		10⅝	65½
15⅛	12	62¼	15⅝	11¼	62¼	16⅛	10½	61¾	16⅝	9⅞	62
	12⅛	62¾		11⅜	63		10⅝	62½		10	62¾
	12¼	63½		11½	63¼		10¾	63¼		10⅛	63½
	12⅜	64¼		11⅝	64¼		10⅞	64		10¼	64¼
	12½	64¾		11¾	65		11	64¾		10⅜	65
	12⅝	65½		11⅞	65¾		11⅛	65½		10½	65¾
15¼	11¾	62	15¾	11	61¾	16¼	10⅜	62	16¾	9¾	62
	11⅞	62½		11⅛	62½		10½	62¾		9⅞	62¾
	12	63¼		11¼	63¼		10⅝	63½		10	63½
	12⅛	64		11⅜	64		10¾	64¼		10⅛	64¼
	12¼	64½		11½	64¾		10⅞	65		10¼	65¼
	12⅜	65¼		11⅝	65¼		11	65¾		10⅜	66
15⅜	11⅝	62¼	15⅞	10⅞	62	16⅜	10¼	62¼	16⅞	9⅝	62
	11¾	63		11	62¾		10⅜	63		9¾	63
	11⅞	63½		11⅛	63½		10½	63¾		9⅞	63¾
	12	64¼		11¼	64¼		10⅝	64½		10	64½
	12⅛	65		11⅜	65		10¾	65¼		10⅛	65¼
	12	65½		11½	65¾		10⅞	66		10¼	66¼

THE IMPERIAL BUSHEL TABLE.

Diameter in Inches.	Depth in Inches.	Content in Pints.	Diameter in Inches.	Depth in Inches.	Content in Pints.	Diameter in Inches.	Depth in Inches.	Content in Pints.	Diameter in Inches.	Depth in Inches.	Content in Pints.
17	9½	62¼	17½	8⅞	61½	18	8⅜	61½	18½	7⅞	61
	9⅝	63		9	62½		8½	62½		8	62
	9¾	63¾		9⅛	63¼		8⅝	63¼		8⅛	63
	9⅞	64¾		9¼	64¼		8¾	64¼		8¼	64
	10	65½		9⅜	65		8⅞	65¼		8⅜	65
	10⅛	66¼		9½	66		9	66		8½	66
17⅛	9¼	61½	17⅝	8¾	61½	18⅛	8¼	61½			
	9⅜	62¼		8⅞	62½		8⅜	62¼			
	9½	63¼		9	63¼		8½	63¼			
	0⅝	64		9⅛	64¼		8⅝	64¼			
	9¾	64¾		9¼	65		8¾	65¼			
	9⅞	65½		9⅜	66		8⅞	66			
17¼	9⅛	61½	17¾	8⅝	61½	18¼	8⅛	61¼			
	9¼	62¼		8¾	62½		8¼	62¼			
	9⅜	63¼		8⅞	63¼		8⅜	63¼			
	9½	64		9	64¼		8½	64¼			
	9⅝	65		9⅛	65¼		8⅝	65			
	9¾	65¾		9¼	66		8¾	66			
17⅜	9	61½	17⅞	8½	61½	18⅜	8	61¼			
	9⅛	62½		8⅝	62½		8⅛	62¼			
	9¼	63¼		8¾	63¼		8¼	63			
	9⅜	64¼		8⅞	64¼		8⅜	64			
	9½	65		9	65¼		8½	65			
	9⅝	65¾		9⅛	66		8⅝	66			

TO ASCERTAIN THE WEIGHT OF LIVE CATTLE.

First, see that the animal stands square, then, with a string, take his circumference just behind the shoulder-blade, and measure the feet and inches—this is the *girth.* Then measure from the bone of the tail which plumbs the line with the hinder part of the buttock, and direct the string along the back to the forepart of the shoulder-blade, and this will be the *length.* Then, work the figures thus. —Suppose girth of bullock 6 feet 4 inches, length 5 feet 3 inches, which multiplied together make 33 square superficial feet; and these, multiplied by 23—the number of pounds allowed for each superficial foot of cattle measuring less than *seven* and more than *five* feet in girth—make 759 lbs. When the animal measures less than *nine* and more than *seven* feet in girth, 31 is the number of pounds to be estimated for each superficial foot. And suppose a small animal to measure 2 feet in girth and 2 feet in length, these multiplied together make 4 feet, which, multiplied by *eleven*—the number of pounds allowed for each square foot when cattle measure less than three feet in girth—make 44 lbs. Again, suppose a calf or sheep, &c., to measure 4 feet 6 inches in girth, and 3 feet 9 inches in length, that multiplied together, makes 16 square feet, and these multiplied by 16, the number of pounds allowed for cattle measuring less than 5 and more than 3 feet in girth, make 256 lbs. The dimensions in girth and length of the back of cattle, sheep, calves and hogs, taken this way, are as exact as is at all necessary for common computation or valuation of stock, and will answer to the four quarters of the animal, sinking the offal. A deduction must be made for animals half fat, of one pound in twenty from those that are fat; and for a cow that has had calves, one pound must be allowed, in addition to the one for not being fat, upon every twenty.

THE END.

www.ingramcontent.com/pod-product-compliance
Lightning Source LLC
LaVergne TN
LVHW021412110826
845150LV00007B/1892

* 9 7 8 1 4 2 5 5 1 0 6 9 5 *